Synthesis Lectures on Ocean Systems Engineering

Series Editor

Nikolas Xiros, University of New Orleans, New Orleans, LA, USA

The series publishes short books on state-of-the-art research and applications in related and interdependent areas of design, construction, maintenance and operation of marine vessels and structures as well as ocean and oceanic engineering.

Fidaa Karkori

Ship Vibration 3

Noise and Vibration Control for Inhabited Ships Spaces

 Springer

Fidaa Karkori
Southampton, UK

ISSN 2692-4420   ISSN 2692-4471   (electronic)
Synthesis Lectures on Ocean Systems Engineering
ISBN 978-3-031-68077-9   ISBN 978-3-031-68078-6   (eBook)
https://doi.org/10.1007/978-3-031-68078-6

This Springer imprint is published by the registered company Springer Nature Switzerland AG
The registered company address is: Gewerbestrasse 11, 6330 Cham, Switzerland

If disposing of this product, please recycle the paper.

To me, the sea is a continual miracle; the fishes that swim, the rocks, the motion of the waves, the ships with men in them. What stranger miracles are there?

—*Walt Whitman*

For Zakaryah Maximilian

Preface

Working and living onboard vessels impose a series of generally low-frequency mechanical vibrations as well as single-impulse shock loads on the human body. Also, exposure to noise is characteristic aboard vessels. Low-frequency vibrations are created by vessel motions, which are produced by the various sea states in conjunction with vessel speed and point of sail. These motions can result in motion sickness, body instability, interruptions of task performance, sleep interruption and fatigue, increased health risk aggravated by shock loads due to slam, and reduced human efficiency.

Higher frequency vibration influences comfort and is often associated with rotating machinery. The imposition of higher frequency vibrations (about 1 to 80 Hz) induces corresponding motions and forces within the human body creating discomfort and reduced human efficiency. With regard to noise, the above can similarly affect exposed humans, notably with sleep interruption and resulting fatigue, discomfort, and reduced efficiency. Also of concern are transient and permanent hearing loss, masking of audible signals, and interruption of speech communication.

The concerns related to levels and characteristics of noise and vibration are covered in this and the first book related to habitability on ships and offshore structures. To be granted any of the associated habitability notations, specific noise and vibration criteria must be met. Ship designers in pursuit of these notations have requested guidance on how to control levels of noise and vibration in inhabited spaces. As a result, this guide has been written. The information presented in this book is intended for guidance only to support vessel designers and operators in controlling vessel noise and vibration in the general case, and more specifically in meeting the requirements of classification society habitability rules. Adherence to some or all of the guidance in this book affords no guarantee that a habitability notation will be granted.

Southampton, UK   Fidaa Karkori
2024

Acknowledgements It is with inestimable gratitude that I extend my greatest thanks to Dr. Dieter Merkle and Prasanna Kumar Narayanasamy, both of Springer Nature, for their support and guidance in the writing of this book and to my husband and fellow author, Alexander Olsen, for his insights, experience, and patience.

The original version of the book has been revised. A correction to this book can be found at https://doi.org/10.1007/978-3-031-68078-6_12

Contents

Abbreviations and Acronyms

A/V	Audio/visual
AB	Airborne
CAD	Computer aided design
CFD	Computational fluid dynamic
FEA	Finite element analysis
HPU	Hydraulic power unit
HTL	High transmission loss
HVAC	Heating, ventilation, and air conditioning
ILO	International Labour Organisation
ISO	International Standards Organisation
MLC 2006	Maritime Labour Convention, 2006
NIST	National Institute of Standards and Technology
PSB	Structure-borne
QA	Quality assurance
R&D	Research and development
rms	Root-mean-squared
SOLAS	International Convention for the Safety of Life at Sea
SSB	Secondary structure-borne
TL	Transmission loss

List of Figures

List of Tables

General

1

1.1 Introduction

Work-related hearing loss and vibration-related health effects are critical workplace safety and health issues. Seafarers may experience motion sickness, body instability, fatigue, and noise-induced hearing loss, one of the most common occupational diseases. Adverse/ improper noise and vibration levels can also cause speech interference, mask warning signals, interfere with concentration and thought processes, disrupt sleep, and create harmful living and working conditions. If designed appropriately, however, vessels complying with proper noise and vibration levels can provide an environment for improved crew performance, safety, comfort, and communication, and have an overall positive psychological effect on seafarers.

1.2 Addressing Noise and Vibration on Ships

The recommended way to meet noise and vibration criteria is to undertake noise and vibration analyses very early in the design process and apply appropriate controls to mitigate areas of potential concern. Noise and vibration analysis includes:

- Identifying sources of noise and vibration,
- Modelling noise and vibration within the vessel,
- Calculation of exciting forces (frequency and amplitude),
- Location of force application and response of the vessel structure in the positions of interest,
- Modelling the source-path-receiver phenomenon, and

© The Author(s), under exclusive license to Springer Nature Switzerland AG 2025
F. Karkori, *Ship Vibration 3*, Synthesis Lectures on Ocean Systems Engineering,
https://doi.org/10.1007/978-3-031-68078-6_1

- Using this information to review the existing design for opportunities to improve noise and vibration levels.

Therefore, this analysis requires methods which can use design information as input data and calculate the expected noise and vibration levels at positions of interest. It is suggested that the exciting forces should be determined for conditions when the propulsion machinery runs at not less than 80 percent maximum continuous rating (MCR) for noise and between 80 and 105% for structural vibration analyses.

1.2.1   New Industry Requirements

There are several industry requirements relating to noise exposure levels for seafarers. These include the IMO Code on Noise and the International Labour Organisation's (ILO) *Maritime Labour Convention, 2006*.

1.2.2   IMO Code on Noise

On 1 July 2014, the *International Convention for the Safety of Life at Sea* (SOLAS) was amended to make the "*Code on Noise Levels Onboard Ships*" (the IMO Code on Noise/ Noise Code) mandatory for new vessels.

The *IMO Code on Noise Levels Onboard Ships* (IMO Resolution MSC.337(91)) is now in force and includes requirements limiting noise for certain types of vessels in excess of 1,600 gross tons. The IMO Code on Noise also includes noise from the operation of thrusters and from noise in port that is related to cargo operations. There are also requirements for acoustic insulation. This Resolution sets out mandatory maximum noise level limits for machinery spaces, control rooms, workshops, accommodation and other spaces on board ships. The IMO Code on Noise supersedes the previous non-mandatory Code, adopted in 1981 by resolution A.468 (XII).

The purpose of the new IMO Code on Noise is to provide standards on preventing noise levels hazardous to human health and reduce seafarers' exposure to such noise levels. It gives consideration to the need for communication, the ability to hear audible alarms, the protection of seafarers from possible noise-induced hearing loss, and the provision of an acceptable degree of comfort during rest hours.

1.2.3   Ilo Mlc 2006

The International Labour Organisation (ILO) *Maritime Labour Convention* (MLC), 2006 was ratified and came into force in August of 2013 by Port States having adopted the

Convention. Within the maritime sphere, the ILO provides legal instruments aimed at protecting and improving seafarers' working and living conditions. The Convention provides the, as yet, most comprehensive Code regarding seafarers' rights, and the obligations of Flag States and shipowners with respect to these rights. The Convention incorporates the fundamental principles of many ILO Conventions and brings together and updates 68 existing ILO instruments (Conventions and Recommendations) into one document.

The MLC has Regulations, both objective and subjective, relating to whole-body vibration and noise levels aboard vessels. Examples are listed below:

(1) Regulation A3.1.6(h) states: "accommodation and recreational and catering facilities shall meet the requirements in Regulation 4.3, and the related provisions in the IMO Code on Noise, on health and safety protection and accident prevention, with respect to preventing the risk of exposure to hazardous levels of noise and vibration and other ambient factors and chemicals on board ships, and to provide an acceptable occupational and on-board living environment for seafarers",

(2) MLC Regulation A3.1.6(h) calls out Regulation 4.3. Below are listed the physical design and arrangement related aspects of Regulation 4.3 including:

    (a) A4.3.1(b) "reasonable precautions to prevent occupational accidents, injuries and diseases on board ship, including measures to reduce and prevent the risk of exposure to harmful levels of ambient factors and chemicals as well as the risk of injury or disease that may arise from the use of equipment and machinery onboard ships."

       Note: Ambient factors refer to improper levels of vibration, noise, lighting, and indoor climatic qualities.

    (b) A4.3.2(a) "take account of relevant international instruments dealing with occupational safety and health protection in general and with specific risks, and address all matters relevant to the prevention of occupational accidents, injuries and diseases that may be applicable to the work of seafarers and particularly those which are specific to maritime employment;"

       Note: Occupational disease refers to disorders such as noise-induced hearing loss, tinnitus, and musculoskeletal injuries/disorders (e.g., lower and upper back and neck issues).

    (c) A4.3.4 "Compliance with the requirements of applicable international instruments on the acceptable levels of exposure to workplace hazards on board ships and on the development and implementation of ships' occupational safety and health policies and programmes shall be considered as meeting the requirements of this Convention."

The more common international instruments related to the "ambient factors" of noise and vibration are the new IMO Code on Noise and ISO 6954: 2000, Mechanical vibration – Guidelines for the measurement, reporting and evaluation of vibration with regard to habitability on passenger and merchant ships.

The IMO Code on Noise and the ISO 6954 vibration criteria should be incorporated into existing classification society suites rules relating to the habitability standards for ships and offshore installations, in addition to each classification society's guidance pertaining to achieving compliance with the ILO MLC, 2006 Title 3 Requirements, and the associated MLC ACCOM notation (or equivalent). These notations support the demonstration of the vessel compliance with the physical design and arrangement requirements, including acceptable conditions of onboard ambient environmental factors of MLC Title 3. It should be noted for completeness that the classification society guides are unlikely to address the procedural or management system requirements required by the ILO MLC, therefore further instruction may be required.

1.3   Application and Scope

The guidance provided in this book is intended to assist the marine community (shipyards, designers, regulators, and owners) in addressing noise and vibration issues and important design parameters. The information throughout the following chapters is recommendatory and should be strongly considered when designing any new vessel in order to provide the safest and most productive working environment for seafarers. The cost of fixing noise or vibration issues can be as much as ten times more expensive after construction than if incorporated into the design from the preliminary design stage. Therefore, careful consideration should be given to designing in noise and vibration reduction elements.

This guidance is intended to provide a basic understanding and overview of the critical factors controlling noise and low-frequency vibrations onboard vessels and to gain better understanding of the important concepts of noise and vibration, design parameters, terminology, analysis methods, acoustic and vibration treatments, and other important data to consider when dealing with noise and vibration.

Moreover, this book also includes basic terminology and definitions of acoustic and vibration terms, a description of noise and vibration generating mechanisms, specifics on noise and vibration sources in ships, information regarding noise and vibration transmission through the hull structure, an overview of methods used for noise and vibration analysis, and noise and vibration control during the design stage or during modernization. With the proper understanding of how noise and vibration is generated, controlled and measured, vessels can be more easily designed to reduce noise and vibration in a straightforward, economic, and optimal manner, thus improving seafarer levels of habitability, safety, and task performance.

The main philosophy of the guidance provided throughout this book is that noise and vibration issues can be identified and resolved through proper attention to the vessel's acoustic and vibratory design including program planning, analysis, treatment selection, construction quality assurance (QA), and verification testing.

Vibration

2

2.1 Introduction

Only costly noise and vibration countermeasures remain after the keel laydown of a ship.

(ISSC 2012)

Working and/or living onboard vessels can impose a series of low- and high-frequency mechanical vibrations, as well as single-impulse shock loads on the human body (whole-body vibration). Low frequency vibrations are also imposed by vessel motions, which are produced by the various sea states in conjunction with vessel speed. Among other things, these motions can result in increased health risks aggravated by shock loads induced by vessel slamming. Vessel slamming may be caused by dynamic impact loads being exerted on the vessel's bottom or bow flare due to vessel size, speed, and wave conditions.

High-frequency vibration is often caused by rotating machinery. The imposition of higher frequency vibrations induces corresponding motions and forces within the human body (whole-body vibration), creating discomfort and possibly resulting in degraded performance and health (Griffin, 1990) *Handbook of Human Vibration*. For the ambient environment related to whole-body vibration, the frequency range of interest is from 1 to 80 Hz.

2.2 Scope

This chapter contains information regarding the proper design and consideration of vibration reduction elements. The design goals are to reduce vibrations to a safe level, thus improving habitability and seafarer task performance in different environments on ships. This chapter contains the following subsections: 3. presents an overview of shipboard

F. Karkori, *Ship Vibration 3*, Synthesis Lectures on Ocean Systems Engineering, https://doi.org/10.1007/978-3-031-68078-6_2

vibration; 4. describes the main sources of vibration; and 5. discusses hull and structure vibration response.

2.3   Overview of Shipboard Vibration

2.3.1   Elastic Vibration

One of the challenges in the design of modern ships is the avoidance of excessive elastic vibration of the hull structures in response to external or internal forces. Such vibration may cause discomfort and interfere with performance of crew duties. Severe vibration may lead to structural damage and negatively influence operation of mechanical and electrical equipment on board.

Elastic vibration can be excited in the form of vertical and horizontal bending, torsional and axial modes of elastic structures of the whole hull, as well as in the form of local vibration of sub-structures and components. Typically, propellers excite the most significant vibration. Flexural vibration can also be excited by forces from reciprocating, and to a lesser degree, rotational machinery, and by external forces of sea waves impacting a vessel.

2.3.2   Vibration Study and Source Design

A typical vibration study can be divided into two parts, analysis of dynamic forces exciting the hull and the response of the hull and its substructures. Looking at the sources of vibration, it is easiest to start at heavy pieces of mechanical equipment and work 'backwards'. It is important to remember that vibration does not really pass through air; it travels along 'things' and the routes that it follows are as important as the vibrations themselves. The most basic forces of interest are those generated by engines, machinery, propellers, and shafts. These forces vary with engine load, the speed and draft of the ship and with environmental conditions. Other common sources of vibration include gears, screws, hulls, thrusters, fans, compressors, pumps, pipes, and valves. It is assumed that shafting, torsional, whirling, lateral, and axial vibration analyses are carried out as part of the standard design, either by the equipment vendor or a third party. This is especially critical when either or both of the coupled equipment items are isolation mounted and require a flexible shaft coupling to take up their expected relative movements.

Selecting low vibration machinery and minimising propulsor/thruster excitation are the better approaches to avoiding excessive vibration levels. Foundations for shaft bearings should be stiff enough to prevent excessive low frequency vibration. The strategy of selecting low vibration machinery is so critical due to a limited or restricted ability to change the hull size and shape later in the design.

2.4 Sources of Vibration

2.4.1 Machinery Excitation

Unbalanced or misaligned machinery, particularly propulsion machinery such as the main propulsion engines, is the major source of excessive vibration and can develop excitation forces in the frequency range of interest—both for the equipment and structure in the vicinity of the machinery. These frequencies of excitation match either the rotation rate, or twice the rotation rate.

2.4.1.1 Diesel Engines

Diesel engines, whether low, medium or high speed—two or four stroke, generate significant force and moment-coupled vibration excitation. The diesel firing forces, which depend on rotation rate and number of cylinders, are impulsive in nature, which causes components of vibration at many harmonics of the firing frequency. Low-speed propulsion diesel engines are typically more of a concern than high-speed diesel engines because their low excitation frequencies are more likely to lie in the range of a hull natural frequency.

As a result, diesel engines, and other reciprocating machinery, may develop unbalanced forces and moments, which are sufficient to create excessive ship vibration. These forces/moments may excite the machinery itself, engine or other machinery foundations, the hull girder, propeller shaft or other structures within the vessel.

The excitation of a reciprocating engine can be divided into two parts:

- Unbalanced inertial, and
- Firing forces/moments.

The unbalanced inertial forces are associated with the rotating or reciprocal masses. The frequencies of resulting forces are multiples of shaft rotation rate.

A 4-cycle engine has twice as many 'firing' harmonics as a 2-cycle engine. Analytical calculations of diesel engine generated forces are not easily undertaken and should be left to the diesel engine vendor. When possible, it is preferable to use experimental data taken on actual diesel engine foundations for a vibration study. Force or vibration levels may be used as input data for response calculation involving structural modelling of the foundation and portions of neighbouring vessel structure.

2.4.1.2 Other Mechanical Sources of Vibration

Other machinery-related concerns are reciprocating refrigeration, air compressors, eccentric rotors, bent shafts, mechanical looseness, rotor rub, bearing problems (whirling, clearances, etc.), and pumps.

2.4.2   Propulsion and Shafting Excitation

The basic design purpose of the propulsors is to generate steady thrust to the vessel. In addition, this propulsor may generate undesired fluctuating dynamic forces and moments due to its operation in a nonuniform wake and due to passage of the blades close to hull and appendages.

2.4.2.1   Hull Pressure Versus Bearing Forces

In addition to propeller induced unbalanced forces, there are two other types of fluctuating forces. These are hull pressure forces and bearing forces. The hull pressure originates from the hydro-acoustic pressure variations caused by the passage of a propeller through non-uniform inflow or wake, and the amplitude of excitations depends on the uniformity of the axial and tangential wake profile. These hull pressure forces are affected by propeller-hull clearance in vertical and horizontal directions, by blade loading, and by changes in the local pressure field around the blades. Given a realistic hull form, it is impossible to achieve a uniform inflow.

However, the design should strive for as clean an inflow as possible.

Bearing forces are caused by fluctuating forces on the blade during a propeller rotation which generate both vertical and horizontal forces on the shaft. These vertical and horizontal forces produce lateral and axial forces and moments on the support bearings and thrust bearing. The frequencies of these forces and moments are the same as for the hull pressure, shaft rate, blade rate, and multiples. The thrust bearing forces provide an excitation to the propulsion system in the longitudinal direction; the fluctuating torque produces shaft torsional vibration. The blade frequency vertical bearing force, when combined with same frequency vertical hull pressures, provides the overall vertical excitation of the hull in the vertical direction. Similarly, the combined horizontal forces excite the hull in the horizontal direction. Dynamic unbalance of any component of the system (propeller, shaft, coupling, gear, and engine) will increase the resulting hull vibration levels.

The decision on the number of blades for the propeller is expected early in the design process. The frequency of propeller blade rate excitations depends on the shaft rotation rate and the number of blades. This applies for all types of screw propellers and impellers, including conventional, controllable pitch, ducted, nozzles, thruster propellers, and pods.

2.4.2.2   Cavitation

Where there is the occurrence of cavitation (the sudden formation and collapse of low-pressure bubbles in liquids by means of mechanical forces, such as those resulting from rotation of a propeller), there is a significant increase in the hull pressure forces. Again, the cavitation performance of the propeller is related to the amplitudes of the blade rate and blade harmonic vibration, and also the broad-band vibration excitations covering a wider frequency range. The cavitation characteristics of the propeller along with the wake interaction and the number of propeller blades with the range of shaft rotation rates all determine the amplitude and frequency of propeller vibration. The resulting hull pressures may be calculated during the design stage based on parameters such as propeller rpm, clearance, blade form, and hull form.

2.4.2.3 Thrusters

Another significant vibration source is the bow or stern thruster(s). These units can be located relatively close to accommodations and can cause high vibration (and noise). As discussed above for the prime propulsor, inflow and propeller design control the induced vibration levels. It is assumed that shafting, torsional, whirling, lateral, and axial vibration analyses are carried out as part of the standard design, either by the equipment vendor or a third party. This is especially critical when either or both of the coupled equipment items are isolation mounted and require a flexible shaft coupling to take up their expected relative movements.

2.4.3 Hydrodynamic Excitation

Hydrodynamic vibration excitation is a flow-related phenomena within the frequency range of 1 to 80 Hz that indicates whole body vibration. The sources of this vibration are generally hull shape, hull appendages, and openings, tunnels or chests subjected to interactions with flow. Vibration excitation induced by hydrodynamic sources is typically unintentional and should be avoided. Some examples of hydrodynamic sources of whole-body vibration excitation includes:

- Pressure pulsations on the hull due to the propulsor,
- Vortex shedding alternating lift forces on underwater appendages,
- Vibrations due to flow over an opening, moonpool, cavity, or tunnel, and
- Wave slap, such as waves acting on shell plate of a flat underbody near the stern or waves against chines.

2.5   Hull and Structure Vibration Response

2.5.1   Global Response

For the global response, one can consider the hull as a beam (of varying stiffness) floating in water and vibrating in a 'free-free' mode. As mentioned before, propulsors and engine excitation create moments and forces which can induce vertical, horizontal transverse, longitudinal, and torsional vibration of a hull, which acts as a whole beam. If the excitation frequency from a particular force is close to or coincides with a natural frequency of the vertical, horizontal, longitudinal, or torsional mode then the whole-body vibration may be significant. For vertical and horizontal vibration, the first two to three modes of vibration are of greatest interest. For hull torsional vibration, only the first mode (one node) may be important. The torsional mode may be coupled with bending modes. The lowest natural frequency of the whole body for commercial ships is typically below 10 Hz.

The normal practice is to calculate a few natural frequencies of the hull, once information about the hull structure and weight distribution becomes available. The main goal of this calculation is to check how close the excitation frequencies come to the computed natural hull frequencies. As a design rule of thumb, forcing frequencies with the machinery and propulsor at maximum continuous speed should differ by 20–30% from the hull mode natural frequencies. This ratio depends on the sophistication and accuracy of the hull modelling process.

Note: *The hull girder natural frequencies can rarely be determined with great precision in preliminary design. It is recommended to include bands of uncertainty around the frequencies of the estimated hull modes.*

Furthermore, the response of the hull will vary with load/ballast conditions and weight distribution along the hull. In a ballast condition, the vibration amplitude may be higher or lower than when lightly-loaded.

An increase in vibration in shallow water has also been observed, with depths varying based on salinity, temperature, and size of the vessel. It is useful to keep these ranges of uncertainty or bands of variability in the documentation of the project for future reference. Empirical and analytical methods for "global" vibration calculations are discussed in Chap. 3, Sect. 1.3: "Vibration calculation by empirical methods".

The hydrodynamic inertia of the underwater body of the hull must be considered in the global hull girder analysis. This effect is due to hydrodynamic pressure reacting against the shell plating and is generally referred to as "added mass". Errors in the determination of the hull global natural frequencies may result if added mass effects are not included. Natural frequencies for the wetted hull are lower in frequency compared to the vibration modes of the structure in a vacuum. For vessels with significant variable displacement, the added mass effect should be considered over the range of drafts for normal operations.

ISO 20283–2 Mechanical vibration—Measurement of vibration on ships, Part 2 Measurement of structural vibration provides a standard guideline for the measurement of global ship vibration. The standard provides recommended methods to obtain an overall picture of the vibration behaviour of the vessel by measurement of natural frequencies and vibration responses related to the hull operating deflection shapes. This information can be valuable to naval architects and structural designers as experimental verification for computations of the global hull modes.

2.5.2   Local Response

Local structures may exhibit a vibration response independent of the 'whole' body response. These substructures, like decks, machinery platforms, shafting systems, or the superstructure, of sufficient mass and flexibility need also be considered in the vibration analysis process. These substructures have their own natural frequencies, which again should not coincide with primary excitation frequencies. The natural frequencies of these substructures are usually higher than that of the whole hull. The frequency depends on the structural stiffness, mass distribution, and on the method of attachment of substructure to the main structure (boundary condition).

The island-type deckhouse may be excited in different ways. Longitudinal vibration of the upper part of a deckhouse may be induced by shear or bending deflections, by support deflections, or longitudinal vibration of a hull. Similar reasons may generate deckhouse vibration in the transverse direction.

Local structural elements like plates, bulkheads, parts of the deck, shafting systems, and machinery platforms may have natural frequencies which are close to the excitation frequencies. There are no reliable empirical methods to calculate natural frequencies of these substructures. Massive equipment installed on platforms may have a large influence on the resulting natural frequency. Local vibration responses become more apparent when the vessel is subjected to significant broad-band excitation energy such as from cavitation of propellers and thrusters, frequent wave interaction/slamming, and during icebreaking. Analytical methods to calculate local vibration are discussed Chap. 3, Sect. 1.3: "Vibration calculation by empirical methods".

Shipboard Vibration Modelling

3

3.1 Introduction

There are three approaches to developing vibration modelling. The most accurate and time-consuming is to build a finite element analysis (FEA) model. The second approach would involve an empirical analysis.

Simplified methods exist to determine the vibrations of 'free-free' beams, framed panels, cantilever structures, etc. These methods are typically less accurate than those using modern FEA techniques. The final method is to use historical data from a close prototype vessel and account for differences in excitation frequencies and machinery. This type of scaling may also be risky but can be used as a first-cut to indicate likely problem areas.

With any of the approaches listed above, the first objective is to predict the structural resonances. If these are sufficiently clear of any of the exciting frequencies, it may not be necessary to carry out a 'forced' response analysis. If a coincidence is likely, a 'forced' analysis can be used to determine if the response is within the allowable limits.

3.1.1 Campbell Charts

Campbell charts are typically used to compare the predicted structural vibration modes to the excitation frequencies. These charts are plotted as (constant) mode frequencies and (variable) excitation frequencies versus operating speed (as a function of either shaft or propeller rotation rate (rpm). Where the excitation frequency curve crosses a hull mode potential resonances can occur. If a reasonable solution is not found to prevent a resonant response, it may be necessary to bar or exclude operation of the equipment or vessel over an operating range where the resonance is present.

© The Author(s), under exclusive license to Springer Nature Switzerland AG 2025
F. Karkori, *Ship Vibration 3*, Synthesis Lectures on Ocean Systems Engineering,
https://doi.org/10.1007/978-3-031-68078-6_3

3.1.2   Finite Element Analysis

Computer-aided calculation methods have greatly expanded the possibility to accurately calculate the dynamic behaviour of large structures such as seagoing vessels. The primary tool currently in use is finite element analysis (FEA), requiring software, qualified analysis personnel, and necessary time for modelling. Using FEA, the whole structure is divided into a large number of small and simple elements: plates, beams, masses, etc. These elements are connected and all together form the model of a realistic structure, refer to Fig. 3.1, "Example of FEA model of a cargo vessel". Book 1 in this series discusses the application of finite element analysis. However, it is recommended to validate the finite element analysis calculations by comparison to model tests and past projects, if possible. This is due to the fact that finite element analyses include various approximations and underlying assumptions in the representation of the physical system and margins of uncertainty.

3.1.2.1   Effects of Water Loading
As previously mentioned in Chap. 2, it is important to include the water-loading effects in the modelling process. The added-mass of the entrained water significantly lowers frequency and response of the hull. Algorithms for simple element vibrational analysis are fairly well established. Special FEA software combines the equations of dynamic behaviour of each element into a system of equations, resulting in a vibration description of the whole system.

3.1.2.2   'Coarse' and 'Fine' Mesh Models
Within FEA, either a 'coarse' mesh model or a 'fine' mesh model can be developed. The coarse mesh model is typically more appropriate for a 'whole' body analysis; the fine mesh model for vibration analysis of 'local' structures. Depending on the vessel's size, the number of elements needed to perform a 'coarse' mesh model is in the tens of thousands. A 'fine mesh' model can increase this by an order of magnitude. The standard FEA task is to calculate the natural frequencies and mode shapes of the hull. Figure 3.2,

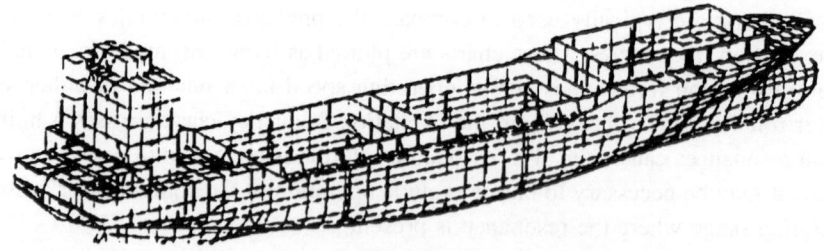

Fig. 3.1 Example of FEA model of a cargo vessel

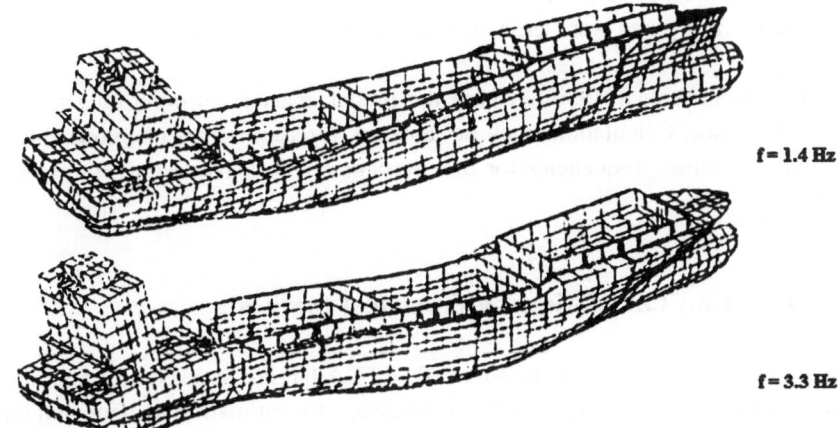

Fig. 3.2 Flexural modes of a cargo vessel

Fig. 3.3 Deckhouse model and bending modes

"Flexural modes of a cargo vessel", shows the torsional (1.4 Hz) and vertical bending mode (3.3 Hz).

3.1.2.3 "Local" Structures

FEA can also successfully be used for the calculation of 'local' structures, such as a deckhouse. Figure 3.3, "Deckhouse model and bending modes", demonstrates a fragment of vessel model for an aft deckhouse. Two bending modes are shown in the sketch.

3.1.2.4 Finding Vibration Response

Knowledge of natural frequencies and mode shape is invaluable information, and in principle gives an opportunity to avoid resonances (coincidence, natural, and exciting frequencies). However, this solution does not quantify the amplitude of vibration. To find the actual vibration response, the following additional information is required:

- Amplitude, frequency, and location of external forces, and
- Loss factors of dry and wet elements as a function of frequency.

The forced vibration analysis often uses the modal analysis as input data to calculate the forced response. Calculations may be done on different ranges of frequencies which correspond to exciting frequencies for that particular vessel, based on various operating conditions.

3.1.3   Vibration Calculation by Empirical Methods

Empirical methods require fewer resources than FEA but are less accurate and universal. A number of empirical relations have been developed for estimating fundamental hull natural frequencies. Vorus, 1988 and Veritec, 1985 both consider the hull as an unsupported beam floating in water. The empirical approach varies with set ship types: tankers, bulk carriers, or cargo ships. The common parameters for all approaches are length, width, displacement, draft, and distance between the bottom and the main deck. Other major parameters are the cross-sectional moment of inertia and mass of a vessel with entrained water. This approach may be very useful in the early design stage and offers an opportunity to avoid the coincidence of natural and exciting frequencies. For example, it might be possible to change the number of propeller blades or shaft rpm early in the design stage.

Other empirical methods can be used to determine the natural frequencies of single and double framed bulkheads/deck, freestanding stacks, and machinery foundations. There are few empirical methods available to conduct 'forced' analyses. 'Four-pole' impedance models have been used to perform forced analyses of simple systems such as isolation mounted equipment and local machinery foundations.

There also exist empirical methods to compute the hull response to propeller hydrodynamic pressures. This approach has been developed to predict the hull response at blade passage based on a few standardised parameters such a tip clearance, hull plate thickness, and the measured or predicted hydrodynamic pressure associated with the propeller. Few tools exist to predict the low frequency habitability vibrations induced by bow thrusters, whether tunnel or drop-down.

3.2   Design for Low Vibration

3.2.1   General

Vibration control should be part of the design and construction process and should accompany this process from the concept design stage through acceptance trials. A systematic approach to vibration control at the design and contraction phases may include the following components:

- Vibration control plan,
- Hull structure treatment,
- Propeller treatment,
- Mechanical source treatment,
- Response calculation, and
- Corrective measures.

Specific details for an overall procedure for assessing ship vibration can be found in Book 1. The general procedures are provided in Fig. 3.4, "Overall procedure for ship vibration assessment".

3.2.2   Vibration Control Plan

Two approaches can be distinguished for the vibration control plan: management approach and engineering approach.

3.2.2.1   Management Approach

The vibration control process requires clear organisation and management. The shipyard should assign a vibration control manager who works as a liaison between the design group, the acoustic consultant, and the production and quality assurance group. If a noise control plan is also being developed, it may be advantageous to have the same manager oversee both the vibration and noise control plans. Acoustic design is an iterative process; sometimes vibration (and noise) treatments may contradict cost/weight/safety issues. The vibration control manager should resolve contradictions jointly with the project manager, the project engineer, and the acoustic design group or acoustic consultant.

During the design process, when the process is based on an iterative vibration analysis, the following should happen:

- The vibration control manager should be involved in assigning material, machinery, vendors, and sub-contractors,

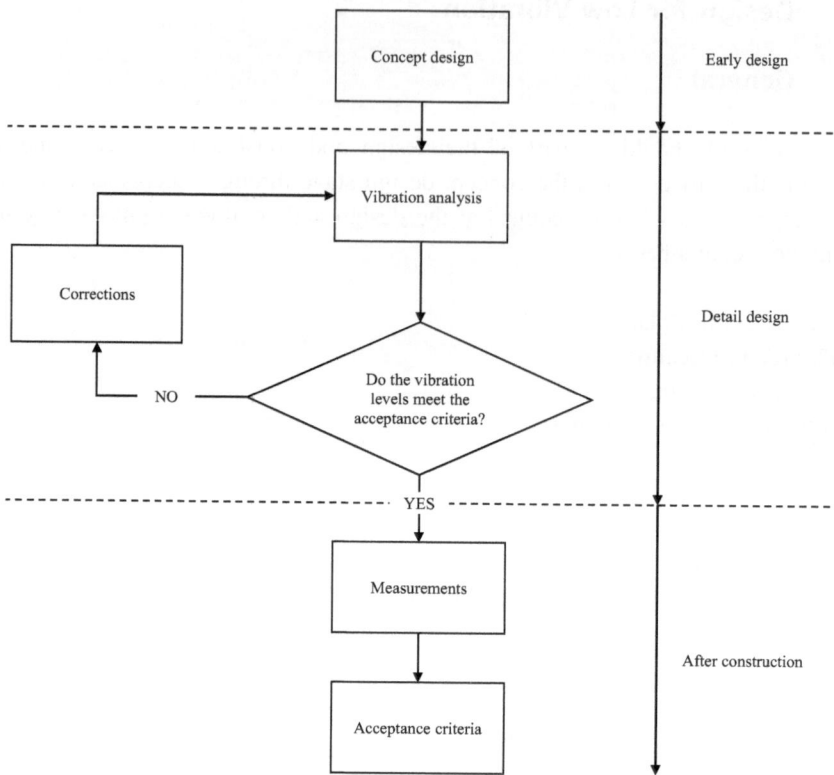

Fig. 3.4 Overall procedure for ship vibration assessment

- Tentative schedules should be developed for internal and external design review meetings, analyses, design deliverables, construction inspections, and trials,
- Particular attention should be paid to major changes in hull structure that have a potential impact on the acoustic performance, and
- The shipyard should develop and submit pertinent schedule/milestone information and identify deliverables per the specification.

3.2.2.2 Engineering Approach

The engineering approach to a vibration control plan may include the following five phases:

- Design review, vibration prediction, primary treatment selection for concept, preliminary and contract design stage,
- Revisions of vibration treatments during detailed design,
- Consideration of non-acoustic impacts,

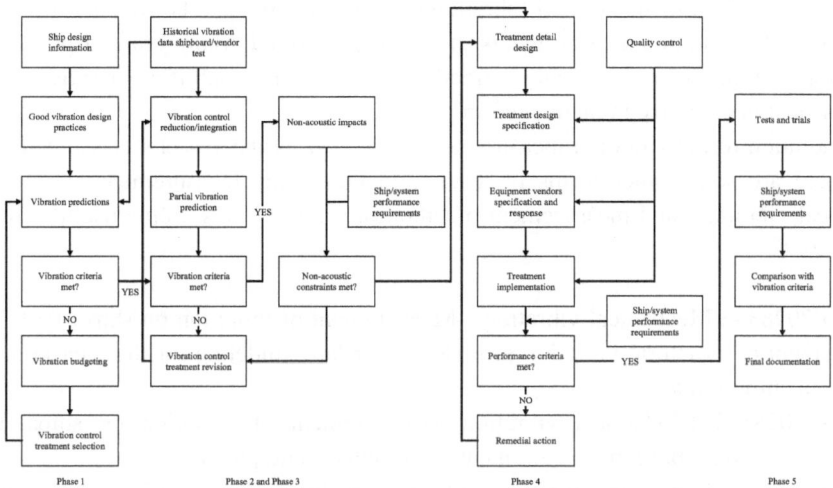

Fig. 3.5 Vibration control plan—engineering flow chart

- Treatment implementation and evaluation, and
- Trial and documentation.

A flow chart is provided in Fig. 3.5, "Vibration control plan—engineering flow chart".

Phase 1: It is assumed that good design practices will be incorporated into the initial ship design stage. Good design practice includes imposing balance and alignment requirements, using relatively low vibration machinery, and providing propeller clearances, etc. Accurate vibration prediction will determine the potential vibration excess area(s), and physical reasons for the excessive vibration levels. As a result of the vibration analyses, the primary vibration treatments can be selected.

Phases 2 and 3: Phases 2 and 3 include trade-off studies, when the proposed vibration treatments are evaluated from a practicality point of view. This includes material and labour cost, weight and space impacts, and meeting any regulatory requirements, such as stress and fatigue. Once the potential contradictions or trade-offs are resolved, all acoustics-related technical solutions should be included in the construction drawings and a final vibration prediction iteration should be performed.

Phase 4: During the construction period, the installers of vibration treatments should be instructed to avoid vibration short circuits in resiliently-mounted systems or missed adhesive for damping treatments. All balance and alignment procedures should be the subject of assurance control.

Phase 5: The final phase in realisation of the vibration control plan is the vibration measurements performed during builder and/or acceptance sea trials. Vibration measurement procedures and reporting should be in accordance with the vessel's classification society rules. If some unexpected vibration or resonance were to be found during the trials, diagnostic vibration measurements should be performed. Many years of practice have demonstrated that the need for any corrective action is very small if an effective vibration control plan is correctly implemented in practice.

In addition to the vibration measurements provided by the vessel's classification society, there are some other forms of testing, including the measurement of hull girder and local vibration, and measurement of vibration source levels. Some useful references include:

- ISO 20283–2 Mechanical vibration—Measurement of vibration on ships, Part 2 Measurement of structural vibration provides a standard guideline for the measurement of global ship vibration,
- ISO 20283–3 Mechanical vibration—Measurement of vibration on ships, Part 3 preinstallation vibration measurement of shipboard equipment,
- ISO 13332 Reciprocating internal combustion engines—Test code for the measurement of structure-borne noise emitted from high-speed and medium-speed reciprocating internal combustion engines measured at the engine feet,
- ISO 8528–9:2017 Reciprocating internal combustion engine driven alternating current generating sets—Part 9: Measurement and evaluation of mechanical vibrations,
- ISO 20816–1:2016 Mechanical vibration—Measurement and evaluation of machine vibration—Part 1: General guidelines,
- ISO 20283–4 Mechanical vibration—Measurement of vibration on ships, Part 4: measurement and evaluation of vibration of the ship propulsion machinery, and
- ISO 20283–5 Mechanical vibration—Measurement of vibration on ships, Part 5 Guidelines for measurement, evaluation and reporting of vibration with regard to habitability on passenger and merchant ships.

3.2.3 Hull Structure Treatment

There is little guidance that will affect the design of the hull, other than providing the proper propeller clearances. Also, a wider and longer superstructure is less risky from vibration point of view. Design of the stern part of a vessel is important because it can provide uniform flow to the propeller, greater clearances, and stiff struts and bottom structure near propulsors.

A foundation for critical vibration sources (e.g., large reciprocating machinery) should have a thick foundation top plate, stiff floors, and local gussets between these two members in way of attachment points. This is true whether the machinery is to be isolation mounted or not.

3.2.4 Propeller Treatment

3.2.4.1 Treatment Methods Using Models

The immediate goal of quiet propeller design is to increase cavitation inception speed. It is therefore a good idea to first require model testing to confirm at what blade rate, and multiples of it, fluctuating pressure is induced. These pulsations can be compared to design rules-of-thumb [2 kPa (0.3 psi)] and used as inputs for any forced vibration analysis.

Furthermore, with modern computational fluid dynamic (CFD) and hydro-acoustic propeller modelling processes, it is possible to 'adapt' the propeller blade form to the operating environment, thereby reducing the vibration (and noise) induced by the propeller. For instance, forward skew on a propeller provides the opportunity to reduce both the induced vibration and noise.

3.2.4.2 Forming the Blade

The production quality of the propeller blades is also an important factor with respect to the induced vibration levels. All blades should be strictly of the same geometry (chord, thickness, pitch, etc.) and to be statically and dynamically balanced. Another method to reduce the fluctuating pressure around the propeller(s) is to reduce the load on the blade peripheral or tip area. Several methods exist to reduce this load. It is possible to reduce blade pitch in tip of the blade while simultaneously increasing the blade area.

3.2.4.3 Propeller Clearance

Propeller clearance is another influential factor. Small clearance, as a rule, leads to excessive vibration levels. Fluctuating pressures decay quickly with distance from the propeller. Thus, larger clearances reduce hull pressures and forces on the bearings. Clearances are usually expressed as a fraction of propeller diameter. Fore and aft clearances are normally measured at the 0.7 radius location, R0.7. Fore clearance is recommended to be not less than $0.25 \times R0.7$. The optimal aft clearance is better determined through model test in a test tank. The tip clearance should be not less than $0.2 \times R0.7$ but is improved by a clearance greater than $0.25 \times R0.7$. The tip clearance is recommended to be a minimum of 20% of the propeller diameter.

3.2.4.4 Bow Thruster Treatment

Bow thruster induced vibration can be minimised by following good acoustic design practices relative to the design of the propeller and the location and placement of the thruster itself (Fisher, 2006). Vendors can also supply resilient supported tunnels (tunnel within a tunnel), bubbly air injectors, and tunnels coated with a decoupling material. The impact in the low frequency range of concern is hard to estimate. Little data exists for measured vibration reduction other than for the higher frequency range.

3.2.4.5 "Add-On" Treatments

"Add-on" treatments that reduce the propeller's impact are typically not simple and not readily implemented. These include air-masker systems that provide 'bubbly-water' along the hull boundary over the propeller, hull compliant coatings on the hull over the propeller, resiliently mounted hull sections, or increased plating and stiffening of the hull over the propeller.

3.2.4.6 Other Treatment Methods

With all conditions equal, waterjets generate lower vibration because the multi-blade impeller in a duct is better balanced and usually has a better inflow than a propeller behind a hull. Azipods and Z-drives located in cleaner flows can have the advantage of low induced-vibration. As for propulsion shaft transmitted vibration, the most effective treatment (other than proper propeller design) is a stiff foundation and proper attention to alignment and balance of the shaft. Regulatory requirements exist for these parameters and should be used.

3.2.5 Mechanical Source Treatment

For vibration caused by unbalanced or misaligned machinery, the 'treatment' in these cases is to correct the mechanical problem. This represents another basic design function often carried out by the equipment vendor, who usually takes responsibility for the proper selection of vibration dampers, torsional couplings, and flexible shaft couplings.

3.2.5.1 Resilient Mounts

If there is no inherent mechanical problem, resilient mounts may provide some abatement. However, isolation mounting provides no vibration reduction below the systems natural frequency and may actually amplify the vibration in the vicinity of the system resonance. Resilient mounts need careful design for two reasons: Vibration levels of the machinery itself are actually higher on resilient mounts than that if the machinery is hard mounted. Another concern is that the excitation frequencies of the machinery should not coincide with isolation system's natural frequencies. To avoid such a coincidence, the choice of the resilient mounts should be accompanied with calculations of natural frequencies of a 'machinery-resilient mount' system (six degrees of freedom calculations). The lower the natural frequency (softer resilient mounts), the lower the vibratory effect will be on the foundation and adjacent structures.

3.2.5.2 Strong Foundation

Designing a stiff foundation structure is the most important approach to prevent excessive machinery induced vibration. The incorporation of a stiff egg-crate like framing and flooring system is preferable and should be a relatively light system of floors in either the longitudinal or transverse direction only. Transverse plating of a foundation should be a continuation of rigid floors; longitudinal plating should be incorporated into longitudinal stiff girders. The structure of thrust bearing foundations is also important, especially for systems with a Cardan shaft (universal joint).

3.2.5.3 Use of Steam and Gas Turbines

Additionally, steam and gas turbines, as rotating machinery, have a distinct advantage relative to diesel engines and reciprocating machinery. Reciprocating equipment of the same nominal power and rpm are known to have significantly different low frequency vibration levels. Thus, it is important to obtain source level information from vendors and choose the vendor providing the lowest possible level. Trades-offs on cost, weight, and long-term maintenance also need to be factored into this decision.

3.2.6 Sea Trials

During sea trials, whole-body vibration measurements should be performed in accordance with appropriate sections of the vessel's classification society rules. The test plan includes selection of transducer locations, measurements taken in specified condition in three axes, analysis of data with a frequency weighting, and combining each axis data to determine weighted multi-axis root-mean-squared (rms) vibration levels and reporting format.

The number and locations for measurements should be chosen to obtain representative data to have a complete picture concerning vibration in manned spaces and work locations. Locations should include the potentially 'worst locations' (e.g., closest to sources or places expected to have maximum vibration based on the design studies). Compartments which represent a single instance of a room type should be included in the test plan (bridge, mess, library, etc.). The measurements should be performed on each deck in a variety of locations including port, starboard, fore, aft, and midship.

Equipment used for vibration measurements should have a National Institute of Standards and Technology (NIST) certificate (or equivalent) and possess enough sensitivity to pick up low background and signal levels which may be expected in some locations. Triaxial accelerometers (measure 3 perpendicular directions) should be used in each location. It is recommended to measure vibration levels in 1/3 octave and narrower band levels between 0.5 and 80 Hz. The third-octave band data will be used as source data for frequency-dependent weighting correction for each direction spectrum. Narrow band data can be used for diagnostic purposes, since the peaks in the measured spectrum can be used to identify machinery and systems that are expected to produce identical tones.

The test report should describe the operating conditions for measurements, including weather, sea state, speed, rpm, water depth under keel, and other pertinent data required by the vessel's classification society rules. The report should show the locations of measurements and overall weighted multi-axis rms value in mm/s^2.

3.3   Implementation

This Subsection addresses the implementation of the vibration control effort onboard the vessel. Once the treatments have been selected and detailed, it is just as important to follow up with good quality assurance (QA) program compliance testing. Furthermore, the vessel's classification society rules will provide detailed requirements on the quality assurance procedures and testing procedures to be submitted.

3.3.1   Quality Assurance

It is absolutely necessary to implement a good quality assurance programme—including a thorough review of drawings and the actual implementation of the treatments onboard the vessel. These steps provide insurance that the time, effort, and cost put into the design effort pays off.

3.3.1.1   Drawing Review

Drawing reviews are an integral part of the vibration control effort, providing needed insurance. As the vibration control treatment designs evolve, drawings and other design information needs to be reviewed by one familiar with the design of treatments to verify that the treatment design adequately reflects vibrational considerations. Results of these reviews need to be documented with appropriate review remarks made on vibration critical drawings. Drawings need to be revised to incorporate any suggested changes to the treatment details.

3.3.1.2   Construction Inspection

The construction of vibration control treatment installations needs to be thoroughly inspected. This includes the inspection of vibration-critical machinery and treatments so that the control treatment installations are not compromised. The inspections should primarily address the fabrication and installation of vibration insulation materials, damping, and other critical treatments. The performance of vibration control treatments is ultimately dependent on the quality of the implementation. Seemingly trivial deviations from the detailed design, or inadvertent errors due to unfamiliarity with vibration control treatment materials and constructions, may compromise vibrational performance.

Comments and action items should document and identifying constructions and installations that are judged to adversely impact vibration performance.

3.3.2   Trials

It is important that proper trials be carried out to determine the effectiveness of the vibration treatment measures. Example vibration test procedures can be found in the vessel's classification society rules. These will include test plans, personnel, conditions, data acquisition and instrumentation, data analysis, test schedule, testing requirements, and test reporting.

3.3.3   Material Selection

Materials selected for installation need to meet regulatory requirements. The most critical of these are fire, smoke, and toxicity. Some materials can be used in machinery spaces but not in accommodations. It is recommended that material vendors with direct experience in the marine industry be approached before considering vendors with only 'industrial' experience. The usual 'installation' standards used by the shipyard and their prior experience in vibration control should also bear consideration in selecting vibration abatement approaches and materials.

3.4   Summary

Human whole-body vibration analysis during the design stage, engineering of vibration control, and participation in construction control and vibration measurements should be performed by companies or personnel with background and experience in vibration control.

Finally, the careful design of propellers, machinery, and hull vibration control may require additional materials and products to reduce habitability vibration levels. These may include but are not limited to damping coatings, resilient mounts for machinery and pipes, and shaft couplings. The common recommendation is to use a company with marine experience that produces or uses products approved by marine classification societies.

Noise

4

4.1 Introduction

Noise is defined as any unwanted or undesirable sound and can be combinations of sounds that are audibly dangerous, distracting, disturbing, or not pleasing to the ear. Sound is a form of energy that radiates from a source in all directions until it gets absorbed or reflected. Noise exposure can be described by the intensity of the noise, its frequency, and its duration.

Noise challenges are especially difficult to address in smaller vessels. The most problematic vessels are those possessing highly-loaded propulsors and thrusters, for example those with towing and dynamic positioning systems. Other noise deficiencies are related to excess noise from the indoor climate systems.

This chapter is intended to provide information for the marine community to gain an overview of the important design parameters and the process that needs to be addressed in order to deliver a 'quiet' vessel that meets safer noise levels.

4.2 Scope

This chapter contains information regarding the proper design and consideration of noise reduction elements. The design goals are to reduce noise to a safer level, thus improving habitability and seafarer task performance in different environments on ships. However, these goals are generally lower than the levels associated with hearing loss. For further information on prevention of hearing loss, see appropriate legislation or regulations, such as *IMO Code on Noise Levels Onboard Ships* (IMO Resolution MSC.337 (91)).

© The Author(s), under exclusive license to Springer Nature Switzerland AG 2025
F. Karkori, *Ship Vibration 3*, Synthesis Lectures on Ocean Systems Engineering,
https://doi.org/10.1007/978-3-031-68078-6_4

4.3   Overview of Shipboard Noise

4.3.1   Sound Pressure Level

When designing for improved levels of habitability, the fundamental parameter of concern is the sound pressure level. Sound is measured by a pressure sensing device, usually a microphone, connected to a sound level meter (SLM) or an acoustic analyser. The units used in reporting the sound pressure level are logarithmic (since the pressure can vary by orders of magnitude) and presented as decibels (dB). The sound pressure is referenced to 20 μPa (0 dB re 20 μPa), which represents the point at which a typical person can just detect sound.

As a standardised quantitative metric for onboard noise assessment, the sound pressure level is measured with a calibrated precision microphone according to the instrumentation requirements established by international standards, including the *IMO Code on Noise*. The range of noise levels which can be measured on the modern vessel is between 30 and 130 dB(A). The lowest may be measured in a compartment on a larger vessel far from the engine room; the highest could be in an engine room supplied with a high-speed diesel engine or gas turbine.

4.3.2   Frequency Range

Generally, the frequency range of interest for human hearing is between 20 Hz and 20,000 Hz. Thus, the meters used to measure noise need to cover a broad spectrum of sound. Noise in this case is defined as 'unwanted sound' or sound associated with operation of the vessel. Since human ears do not respond equally well to different frequencies, an A-weighted filter has been developed. This filter enhances the mid-frequencies between 1,000 and 5,000 Hz and significantly deemphasises the very low and very high frequencies. When the noise from 20 to 20,000 Hz is passed through this A-weighted filter and the sound energy is combined, an 'overall A-weighted' noise value is determined and is denoted as the sound pressure in dB(A). This A-weighted value is commonly called out in noise regulations or requirements and is the applicable metric for the *IMO Code on Noise*. However, sometimes an "equivalent continuous A-weighted sound pressure level" or L_{Aeq} is used. This is basically a time-weighted average over a specific time interval, usually less than 30 s.

Often for analysis or diagnostic purposes the frequency range is subdivided. The most common is 'octave' frequency bands. These bands are identified by the middle frequency of the band. The octave frequency bands of interest in noise control practice are typically: 31, 63, 125, 250, 500, 1000, 2000, 4000 and 8000 Hz.

4.4   Sources of Noise

Each vessel has numerous mechanical, aerodynamic, and hydro-acoustic sources. These noise sources include:

- Propulsors,
- Machinery,
- HVAC and piping systems, and/or
- Wave interaction.

4.4.1   Propulsion Systems

Noise sources may be located outside the vessel or installation. Propellers, thrusters, waterjets, and other hull protrusions fall in a category of 'hydro-acoustic' sources. Excessive noise levels from propulsors are basically connected with their cavitation. Cavitation produces extremely broadband noise along with tonal noise at the blade passage rate (e.g., rotation rate × number of blades). The propulsors generate relatively high sound pressure levels in the water around the blades. This high sound pressure induces vibration on the hull (bottom shell or thruster tunnel), while part of the acoustic energy may travel through the hub and shafts as well. In accordance with the nature of hull excitation, one can consider a propulsor as a source of a pure structure-borne sound. Methods of propulsor evaluation as acoustic sources are discussed in Chap. 3, "Shipboard noise modelling".

4.4.2   Machinery

Typical shipboard noise sources include the main and auxiliary engines, pumps, compressors, fans and other equipment located in a machinery space. These sources generally generate significant noise and vibration, particularly in close proximity to the engine space. The main engine's intake and exhaust systems also often generate high noise levels at on-deck stations and, sometimes, inside the vessel's manned compartments.

4.4.2.1   'Airborne' verse 'Structure-Borne' Noise

The acoustic source may generate 'airborne' noise or 'structure-borne' noise. Noise emanating from the casing of any mechanical source, intake/exhaust, or from fans, is generally considered 'airborne' noise. The acoustic source also generates 'structure-borne' noise via the attachment point to its foundation. The 'airborne' noise spreads through air, while 'structure-borne' noise spreads through the structure of the vessel. Received noise in a compartment results when the 'structure-borne' noise radiates from the structure back to the air.

The best noise models and noise control solutions critically depend on defining and understanding these two discrete sources for the same unit. Thus, each of these components should be analysed with different methods and treated differently. Methods for airborne and structure-borne noise calculations during the noise prediction process are discussed Chap. 3, "Shipboard noise modelling".

4.4.2.2 'Airborne' verse 'Structure-Borne' Paths

As with the source, there are two major transmission paths—airborne and structure-borne. There is also another path, called 'secondary structure-borne', which is a result of the interaction of the two (refer to Fig. 4.1, "Airborne/structure-borne source/path").

Noise path analysis requires information regarding the hull structures, joiner panels, levels and types of insulation, coating structures, machinery location, compartments of interest, their location, and the vessel's general arrangement. Noise path analysis includes analytical consideration of sound waves spreading through the air and the structure, as well as analysis of the structure's ability to insulate, absorb, and radiate sound. Consideration and analysis of all this information can be a labour-consuming part of noise prediction. Fortunately, acoustic CAD models make this an easier and more accurate process.

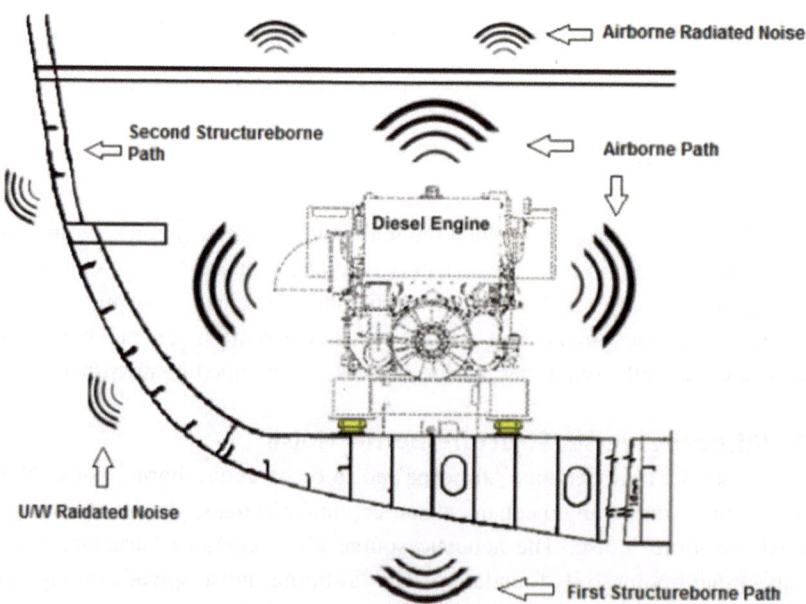

Fig. 4.1 Airborne/structure-borne source/path

4.4.2.3 Noise 'Source' Levels and Octave Bands

Machinery noise and vibration source levels, in general, scale with the equipment power, rpm, and weight. Various empirical methods exist to predict these source levels if measured data from the vendor is not available (Fischer, 1983; 2001). Measured data from the vendor or from similar equipment from a previous project is always preferable to using an empirical approach.

Noise 'source' levels should be measured at a predetermined distance from the unit [usually 1 m (3.3 ft)] and are used for the airborne source characterisation needed for acoustic modelling. It is generally best to determine these levels in octave bands so they can be used as input data for a detailed noise analysis.

Octave band vibration levels taken on the machinery feet should be used as input data for structure-borne noise analysis. Accelerometers are the most practical transducer to measure vibration in the frequency range of interest. As with the sound pressure levels, acceleration vibration levels may be provided in decibels, where the reference is usually 10–6 m/s^2, and reported as 0 dB re 10–6 m/s^2. There are a number of applicable international standards for determining sound power levels for general applications (e.g., ISO 3744:2010—*Acoustics—Determination of Sound Power Levels*). For shipboard equipment, there are no specific international standards, but an example of a relevant national standard is found in ANSI/ASA S12.67—*Pre-installation of Airborne Sound Measurements and Acceptance Criteria of Shipboard Equipment*.

To have accurate 'source' vibration data and to avoid the foundation's influence on acceleration levels, machinery should be installed on the soft resilient mounts when measured on a test bed at the factory. For information on standard guidelines for procedures to assess structure-borne noise source levels as part of factory acceptance tests, refer to ISO 20283–3 *Mechanical vibration—Measurement of Vibration on Ships, Part 3 Pre-installation Vibration Measurement of Shipboard Equipment*.

4.4.3 Heating, Ventilation, and Air Conditioning (HVAC) Systems

As a distributed system, the heating, ventilation, and air conditioning (HVAC) system may be an important noise contributor to manned compartments and workspaces. Generally, ventilation noise is a combination of aerodynamic (fan and flow regenerated) and mechanical noise. Similarly, ventilation systems can be of concern, particularly on deck or in machinery spaces with the internal equipment secured. Note that more than any other system, HVAC-induced noise can be abated or reduced by following good acoustic design practices (discussed in Chap. 6, "Design for Low Noise Levels"). The general objective is that the HVAC induced noise is at least 5 dB below the compartment's noise limit.

HVAC systems require special consideration as acoustic sources. A fan or compressor is typically the noisiest part of the system. In a room where a fan is located, a fan may be considered as a machinery-type source (airborne and structure-borne noise source).

However, for receiver compartments, the concern is noise generated by a fan that propagates along the duct to the terminal louver. The terminal louver can also be considered as a source of airborne induced flow noise in the room of interest. High flow velocities in the duct will generate considerable flow-noise at junctions. The method for HVAC noise calculation and control are well developed and discussed in Chap. 6, "Design for Low Noise Levels".

4.4.4 Piping

While not as distributed as an HVAC system, piping systems can be an important contributor to noise in compartments near the machinery space, (i.e., the control room). Hydraulic systems, especially those with distributed piping runs, can produce annoying noise at the pump pulsation rate. This normally results when the fluid-borne noise in the piping couples to the vibration in the pipe wall. This pipe wall vibration is transmitted as structure-borne noise at the piping supports. Furthermore, the Hydraulic Power Unit (HPU) itself can be an important contributor to on-deck noise.

Pumps, especially HPU's, require special consideration as acoustic sources. Methods for piping noise calculation and control are well developed and discussed in Chap. 6, "Design for Low Noise Levels". For acoustic pressures in pipes, the method found in ISO 15086 *Hydraulic fluid power—Determination of the Fluid-borne Noise* can be used to evaluate the acoustic pressures transmitted in the piping and fluid systems.

4.4.5 Other Sources

Ice, waves, and inflow interaction with the hull may be additional significant noise sources in the bow and after compartments of some ships and should be evaluated. The more stringent the noise limits to achieve, the more critical these sources become.

Electronic equipment and their cooling systems may also be significant sources of noise in noise-critical compartments, such as the Radio Room and Pilothouse. Audio/visual (A/V) equipment (both installed and portable) is another source that may also need to be addressed.

This chapter explores the 'source-path-receiver' phenomena involved with shipboard noise analysis and control. To properly predict noise and optimise the selection of treatments to meet a certain noise criteria (material discussed in Sect. 2, "Shipboard noise modelling", and Chap. 6, "Design for low noise levels"), it is critical to know and understand:

- The acoustic 'source' level of noise producers,

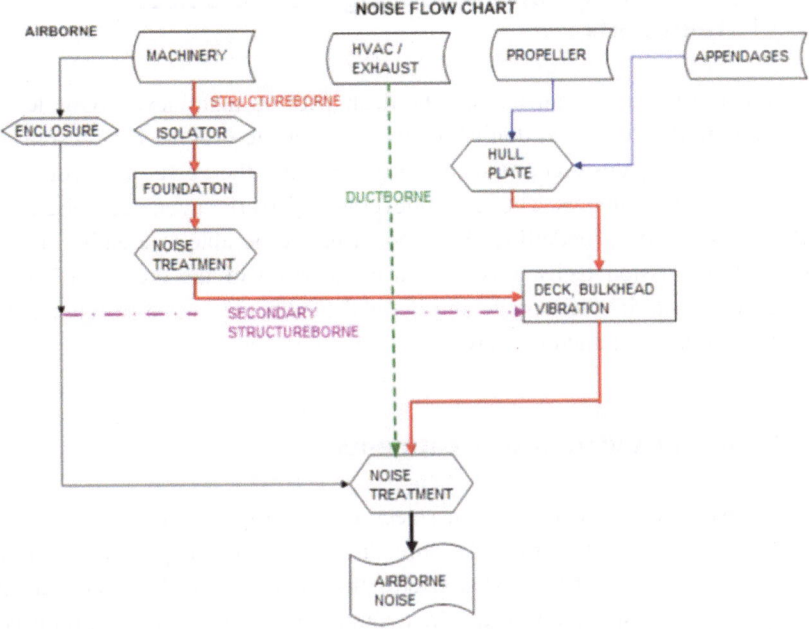

Fig. 4.2 Noise flow chart

- How this energy gets to the receiver via airborne, structure-borne, fluid-borne, or duct-borne paths, and
- How the acoustic characteristics of the receiver space affect the resulting noise levels.

This process is not necessarily simple since all the above parameters may change with frequency and the composition of the vessel between the various sources and multiple receivers. A simple flow chart of this process is provided in Fig. 4.2, "Noise flow chart".

As mentioned previously, the acoustic paths on ships are just as unique as are the sources, including airborne and structure-borne paths and the interaction between them. In addition, there are fluid-borne and duct-borne paths from the piping, propulsors, and HVAC systems. The receiver is a compartment, on-deck station, far-field noise, or the seafarer.

4.5   Acoustic Design and Modelling for the Selection of Treatments

The source-path-receiver parameters can be dealt with on many levels. One level would be to incorporate the "do's and don'ts" of proper acoustic design. This design guidance may assist in avoiding designing potential problems into the vessel and reduce the need for additional noise treatments. The second level is to perform a detailed noise assessment or prediction using noise modelling. This model has to be able to identify critical noise sources and transmission paths. Based on a comparison of the predicted levels versus the desired noise criteria, appropriate and optimal treatments can be selected and applied either to the source, path and/or receiver.

4.5.1   Proper Installation of Treatments

Once the investment has been made in selecting the proper source-path-receiver treatments, it is equally important to make sure these are installed properly. This requires experience and understanding of what makes the treatment work, or conversely, what particular part of the installation makes the treatment fail or become compromised. Without the proper quality assurance (QA), the investment in analysis and design may go for naught. Finally, verification testing is required, and thus verification methods are discussed herein as well.

Seldom is the "as built" sound insulation rating as good as those values measured under controlled laboratory conditions. This is due to flanking paths of sound transmitted above, around or under the rated wall partition, sound transmitted by other paths such as airborne transmission through connected ducts, or due to installation imperfections such a leak between panels, gaps at the perimeter, or cut-outs for wireways and switches. Beyond including appropriate design margins to achieve minimum performance, it is important that design and construction personnel understand these important details to achieve acoustic performance. The noise control plan and the details for noise design should document these requirements and confirm compliance by quality assurance inspections and tests.

Shipboard Noise Modelling

5

5.1 Introduction

This chapter discusses the critical source-path-receiver parameters affecting shipboard noise. These sources are numerous, well-distributed, and this acoustic energy is easily transmitted throughout the vessel due to low inherent losses of a typical metallic hull.

Without an accurate model, it is difficult to select the optimal and appropriate noise control treatments which, if chosen incorrectly, may be excessive or ineffective. Given the usual adverse impact on weight, space, and cost due to adding noise control treatments, it is more cost effective to develop and use a proper acoustic model. This not only saves money, but also labour, and helps to avoid after construction corrections. Retrofitting noise treatments can be up to ten times more expensive compared to incorporating these treatments during the design process. Acoustic models should include the following components:

(1) Source description (e.g., machinery, propulsors, and HVAC),
(2) Acoustic path description (e.g., airborne, structure-borne, and secondary structure-borne paths), and
(3) Receiver space description—including insulation and sheathing.

5.1.1 Machinery Source Description

The machinery source description usually includes:

- Noise and vibration levels (usually in octave bands),
- Size and mass,

© The Author(s), under exclusive license to Springer Nature Switzerland AG 2025
F. Karkori, *Ship Vibration 3*, Synthesis Lectures on Ocean Systems Engineering,
https://doi.org/10.1007/978-3-031-68078-6_5

- Location coordinates, and
- Foundation parameters.

5.1.2 Propulsor Source Description

Propulsor source description usually includes:

- Number of propellers (impellers),
- Number of blades,
- RPM,
- Propeller diameter,
- Clearance between hull and tips of propeller,
- Cavitation inception speed,
- Speed of interest, and
- Other pertinent information.

5.1.3 HVAC Source Description

HVAC source description usually includes:

- Fan parameters (flow rate, power, in-duct sound power levels and pressure),
- Duct parameter (geometry, inside/outside coating, and details of acoustic insulation used in the duct),
- Terminal louver geometry,
- Receiver room sound absorption quality,
- Acoustically lined plenum chambers (if any), and
- Duct attenuators used (if any).

5.1.4 Acoustic Path and Receiver Space Description

Depending on the modelling method, the sound path description should include that part of the ship's structure between the source and receiver or include the whole vessel. When considering the nature of noise and vibration, its spread and distribution through the hull is three-dimensional.

Acoustic energy can spread not only through the shortest path between source and receiver but also through flanking paths. Methods to incorporate these path descriptions into an acoustic model are discussed in the following paragraphs. Additional parameters that need to be considered in the sound path description include hull structure sizes and materials, (damping) loss factors, insulation, and joiner panel parameters.

5.2   Source-Path-Receiver Modelling

As mentioned before, noise is transmitted from a source location to a receiver area over the air media (airborne) and/or through the structure (structure-borne) before it reaches a receiver area of interest. This receiver area can be inside or outside the vessel, in the equipment room, or remote from the equipment.

The airborne source level and airborne paths are the most critical factors affecting noise within a machinery space itself and the compartments directly adjacent. However, structure-borne sources and the structure-borne paths are responsible for carrying the acoustic energy everywhere else on the vessel.

Depending on the level of treatment, secondary structure-borne noise (a combination of the airborne source level and the response of the structure inside the machinery space itself) may also be important in spaces remote from the machinery.

5.2.1   Airborne Noise

The airborne noise source level is the controlling factor in any machinery space. In this case, the received noise level at any position in a source room is the sum of two components: direct noise and a 'reverberant' noise, where the direct noise component prevails in the vicinity of the source and the second (reverberant noise) dominates further from the source. Thus, the direct noise depends on the distance from the machinery to the receiver while the reverberant noise is a result of sound reflection from room boundaries and accounts for the insulation/finish surface of the compartment.

The vessel's classification society rules will provide the minimum requirements for sound insulation between accommodations. The sound insulation rating is determined in a standardised test according to *ISO 717–1 Acoustics—Rating of sound insulation in buildings and of building elements—Part 1: airborne sound insulation* and *ISO 10140–2 Acoustics—Laboratory measurement of sound insulation of building elements—Part 2: Measurement of airborne sound insulation*. The standardised test procedures help account for variability in the as-built construction found on board ships.

5.2.2   Structure-Borne Noise

Structure-borne noise is the result of sound waves radiating from vibrating surfaces (e.g., the deck, bulkheads, deckhead, etc.) within the compartment of interest. This vibration is transmitted by different types of waves, resulting in the structure-borne energy being partially transmitted through obstacles and partially dissipated by conversion to heat and radiation along the path from the source to the receiver. Thus, the greater the number of obstacles along the noise/vibration path, the lower the structure-borne noise will be at the receiver space. Such obstacles include deck/ bulkhead intersections, frames, and other non-uniformities in the structure.

5.2.3   Receiver Spaces

The acoustic characteristics of a receiver space controls the level of the received noise transmitted over any path; whether airborne or structure-borne. Three basic types of receiver spaces should be considered:

- Those with an internal noise source,
- Those without an internal noise source, and
- Those adjacent to a machinery space.

For rooms without an internal noise source that are not adjacent to a machinery space, only the noise radiating from the room's boundaries contribute to the received noise within that space. In this case, two factors are important; vibration levels (such as those of the deck, bulkheads, and deckhead) and the ability of the structure to radiate noise. This ability depends on the plate type and thickness as well as the spacing of the framing. Each of these factors should be quantified in the noise modelling process of maritime vessels.

For compartments adjacent to a machinery space, the airborne noise transmitted through the common interface is critical to the received noise level. This noise, transmitted over the airborne path, depends on the thickness of the interface plating, the insulation, and the finished surface.

5.3   Sname Design Guide and Supplement

The "Design Guide for Shipboard Airborne Noise Control" was issued as SNAME Technical and Research Bulletin No.3–37 in 1983 (Fischer, 1983). This SNAME Guide provides a step-by-step description of a source-path-receiver modelling technique. It is based on empirical prediction techniques gathered from many naval projects. A supplement to this Guide issued in 2001 added new noise calculation algorithms including propeller noise calculation, structure-borne noise paths, and mechanical source updates.

   Calculations taken from these publications can be placed into user-generated computer spreadsheets. Each compartment of interest requires an independent calculation from each influencing source and path.

   Separate sheets are needed to calculate noise in compartments containing a source and in compartments adjacent and distant from a source room. A one-dimensional noise spreading model is used in both the 1983 and 2001 publications. These publications are supported with tables describing acoustic effectiveness of marine hull structures, damping, and absorptive materials.

5.3.1   Analytical Tools

Several noise prediction/analysis software tools currently exist. With these tools, the user can predict the octave band, an overall A-weighted noise level in all rooms of interest, vibration levels of all structural elements of the model (decks, bulkheads), and airborne and structure-borne noise contribution for every room of interest. This type of software can aid the user in the evaluation of noise contributions from every source over every path (including propulsors and inflow which may also be influencing factors) and determine the physical reasons for excessive noise levels, ultimately detailing the optimal way for noise reduction.

5.3.2   Algorithmic verse Modular Models

Noise software packages can be modelled on an algorithmic or modular approach. Algorithmic models are mathematical models which the software uses to determine the source points of noise and provide a realistic representation of the vessel or installation. Modular-based software uses different modules, determined by the industry, in which the modeler is focused. These modules can then be adjusted based on individual customer needs.

5.3.3   Choosing the Correct Software Package

In terms of limitations, when using noise mapping software, 3D mapping is preferred to 2D software, as 2D software does not provide a complete solution. Another consideration in choosing software packages is the type of computer system on which the model will be run.

5.3.4   Finite Element Method

Finite element analysis (FEA) for acoustics can be applied for some predictions, especially in the low frequency range where there are a limited number of acoustic and structural modes to evaluate. Acoustic finite elements are useful for the analysis of duct and silencer predictions where the dimensions of the duct allow for cross modes that limit the applicability of one-dimensional transmission line calculations.

Energy Finite Element Analysis extends the finite element method to higher frequencies using the mesh to predict the spatial distribution of energy in the structure. The method has been implemented for ship structure-borne noise analysis. An advantage of energy finite element analysis is that the finite element structural mesh developed for stress and vibration predictions can be used in a higher frequency range for acoustic predictions.

5.4   Other Modelling Approaches

In some cases, modelling is not absolutely necessary. If a prototype vessel exists that is similar to the vessel of interest in size, general arrangements, and machinery arrangement, the expected noise levels for new vessel may be scaled. If this is done, consideration should be given to the differences in noise and vibration levels of the major sources and the difference in noise treatment effectiveness.

It is important to recognise that any prediction method is an idealised representation. Some prediction methods offer compelling colourised photorealistic geometry of predicted results. However, any deterministic prediction relies on valid and correct input information for the source data, the acoustic characteristics of rooms, and the parameters of transmission paths. It is always necessary to validate calculations with measurements and calculations.

Design for Low Noise Levels

6

6.1 Introduction

As discussed previously, effective treatments exist for abating noise generated by a source, along a particular path, or in the receiver space. Different treatments may be needed to reduce airborne sources, structure-borne sources, airborne paths, structure-borne paths, HVAC induced noise, etc. Each treatment type depends on an understanding of the prevailing airborne or structure-borne noise components. Generic treatments used to abate shipboard noise are discussed in the following sections.

6.2 Noise Control Plan

A noise control plan should be the first thing considered when treating sources of noise and should address both the managerial and engineering approaches to noise control. The vibration aspects are generally considered first in the project. This is because the most efficient design with respect to vibration is the avoidance of resonant responses coincident with major excitations. Normally, the vibration characteristics have first priority for selecting the hull form, propulsor, and machinery. The noise plan then usually develops after the primary characteristics of hull and propulsion are determined. A typical noise control plan should include a systematic approach to noise control from design through verification testing.

© The Author(s), under exclusive license to Springer Nature Switzerland AG 2025   41
F. Karkori, *Ship Vibration 3*, Synthesis Lectures on Ocean Systems Engineering,
https://doi.org/10.1007/978-3-031-68078-6_6

6.2.1   Management Approach

The noise control process requires clear organisation and management. The designers/construction yard should assign a noise control manager who works as a liaison between the design groups, the acoustic consultant, and the production and quality assurance groups.

Acoustic design is an iterative process and sometimes noise treatments may contradict cost, weight, or safety issues. The noise control manager should resolve contradictions jointly with the project manager, project engineers, and the acoustic design group or acoustic consultant.

Based on an iterative noise analysis during the design process, the noise control manager should be involved in assigning material, machinery vendors, and sub-contractors. Tentative schedules should be developed for internal and external design review meetings, analyses, design deliverables, construction inspections, sea trials, and final verification testing. Particular attention should be paid to major changes in hull structure that have a potential impact on the acoustic performance. The construction yard should develop and submit pertinent schedule/milestone information and identify deliverables per the specification.

6.2.2   Engineering Approach

The engineering approach to a noise control plan may include any or all of the following phases:

(1) Design review, noise prediction, primary treatment selection for concept preliminary and contract design stage,
(2) Revisions of noise treatments during detailed design,
(3) Consideration of non-acoustic impacts,
(4) Treatment implementation and evaluation, and
(5) Trial and documentation.

A flow chart is provided in Fig. 6.1, "Noise control plan—engineering flow chart".

Phase 1: For phase 1, it is assumed that good design practice will be incorporated in the initial design stage. Good design practice includes using relatively quiet machinery and propulsors, optimal general arrangements, and placing compartments with less stringent noise criteria closer to major noise sources. Accurate noise prediction can determine the potential noise excess area(s), and physical reasons for the excessive noise levels, and quantify airborne and structure-borne components. As a result of noise analysis, the primary noise treatment can be selected.

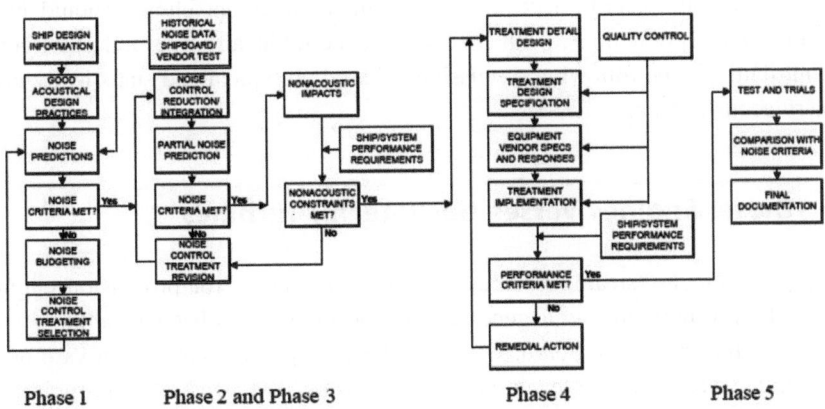

Fig. 6.1   Noise control plan—engineering flow chart

Phases 2 and 3: Phases 2 and 3 may include trade-off studies, when the proposed noise treatments are evaluated from a practicality point of view. This includes material and labour cost, weight and space impacts, and meeting any regulatory requirements, such as fire, smoke, and toxicity. Once potential contradictions or trade-offs are resolved, all acoustics-related technical solutions should be included in construction drawings and final noise prediction iterations are performed.

Phase 4: During the construction period, the installers of noise treatments should be instructed to avoid vibration short circuits in resiliently-mounted systems and floating floors, missed adhesive, super compressed insulation materials, the painting of perforated metal, and so on. All noise treatment procedures should be the subject of assurance control.

Phase 5: The final step (Phase 5) is noise measurements, to be conducted during builder and/or acceptance sea trials. If there is any unexpected noise excess found during the testing, diagnostic noise and vibration measurements should be performed. In practice, experience has shown that the need for corrective action is very small if an effective noise control plan is correctly implemented.

6.3   Machinery and Other Source Treatments

A good rule for noise control is to specify or select equipment with inherently low noise and/or vibration source levels. Equipment with low noise and vibration source levels may eliminate the need for extensive treatments. Typically, there is the trade-off between noise and weight and the cost and operation of the vessel. However, one potential beneficial side effect of using quiet equipment is the reduced maintenance requirements. Inherently quiet equipment is further discussed later in this section.

When treatments become necessary, the optimal treatment choice should be based on accurate noise predictions, with a separation between airborne and structure-borne noise contribution. The following discussion is based on typical and statistically averaged machinery acoustic data.

6.4   Diesel Engines Verses Gas (Steam) Turbines

Diesel engines or gas (steam) turbines are the usual choice for propulsion engines for most vessels. Furthermore, as a general rule, the main propulsion machinery space is the noisiest compartment on board. However, the ability of a construction yard or design office to influence the noise levels alone from propulsion machinery is typically very restricted. Therefore, selection of equipment based on known acoustic characteristics may be more feasible. The objective of this choice is to choose, for a given power, rpm, fuel consumption, mass, size and so on, units that inherently have, or are supplied by the vendor with, lower noise and vibration levels than 'standard' commercial equipment. Considerations to keep in mind when making this decision are:

- Rotating equipment is quieter than reciprocating,
- Drip-proof electric motors are usually quieter than totally enclosed fan-cooled motors, and
- Pumps operating at their most efficient design point are less noisy than those operating off design.

In general, gas turbines are quieter than diesel engines, and produce lower vibration levels due to the inherent differences between rotating and reciprocating machinery. It should be noted, however, that some propulsion diesel engine vendors can deliver quieter units with minimal impact on performance, cost, or weight. However, vendors will sometimes provide optimistic claims that may be misleading, and thus, it is necessary to seek proper validation of vendor low noise claims by obtaining third party noise and vibration data. Verify that the purchase specification provides the technical information, invokes standard measurement procedures and test conditions, addresses the question of guarantees and delivery dates, and considers sources of action for equipment that fails to meet the criteria.

6.4.1   Advantages and Disadvantages of Diesel-Electric Engines

In terms of an overall recommendation for quiet propulsion systems, diesel-electric drive systems have many advantages as the direct diesel drive is through a gear box. In this case, diesel generators are the noisiest part of the electric drive system but are easily installed on resilient mounts such as elastomeric mounts or hybrid spring/elastomeric type mounts.

Another advantage of this type of system is that the electric drive motor is located in the aft section of the vessel, usually far from noise-sensitive areas.

However, direct drive diesel engines with gear boxes are a serious problem for noise control.

Isolation mounting of the diesel engine can help to reduce this noise but would require the use of a very flexible shaft coupling, driving up the system cost significantly. In addition, all the fluid, air, and exhaust connections would also need flexible connections. Also, there may only be a few ways to isolation mount the gearbox, especially if there is no separate thrust bearing. On smaller systems, such as those used on patrol boats and some offshore work boats, it is possible to design an isolation-mounted system for an integral diesel-gear box system, where the isolators actually take the propeller thrust. However, isolation mounting of the prime mover and gearbox becomes less of a concern for waterjet propulsion where there is no propulsion thrust on the shafting.

To help avoid as many of these problems as possible, choosing the right vendor, as emphasised above, becomes a critical decision. There can be up to a 6 dB decrease in noise between normal and high manufacturing accuracy with another 5 dB if extremely high manufacturing accuracy is paired with stringent inspections.

6.5   Acoustic Enclosures and Other Propulsion Noise Treatments

To reduce propulsion noise from gas turbines, the installation of an acoustic enclosure can produce 10 to 20 dB lower airborne noise levels than an equivalent enclosed diesel engine. The effectiveness of this enclosure is the critical feature for noise quieting of gas turbines. For ships equipped with diesel engines, though not the rule, an acoustic enclosure can also be fitted. Such enclosures are more commonly used on diesel generator sets. For installation of the enclosure, there are many diesel engine vendors who supply 'quiet' diesel engines fitted with noise shields.

In general, all propulsion equipment, whether gas turbine or diesel engine, should have an effective silencer or spark-arresting muffler. The diesel engine exhaust system should also have resilient support/mounts if the casing passes through noise sensitive compartments.

Additionally, it is recommended that all diesel generators be resiliently mounted, and gas turbines be isolation mounted. Isolation-mounted systems that are appropriate for the gas propulsor/generator are often recommended by the vendor. Finally, since the unit for diesel generators generally comes on a skid or sub-base, this system is amenable to isolation mounting.

6.5.1   Hard Mounted

If a propulsion diesel engine is hard mounted, structure-borne noise is usually higher than the airborne noise levels caused by the diesel engine in adjacent compartments. The most effective treatment in this case is a resilient mounting system. The alternative to isolation mounting is the extensive application of damping material, floating floors, and isolated joiner panelling.

6.5.2   Resiliently Mounted Engines

If the propulsion engine is resiliently mounted, the contribution of structure-borne and airborne noise is on the same order in adjacent compartments. For noise reduction within the engine room, the use of damping material and sound absorptive insulation should be considered. For remote compartments outside of the engine room, the use of damping materials, floating floors, and heavy joiner panels can help reduce unwanted noise. The compartments requiring these types of treatments should be determined by analysis. Typically, compartments separated by two or three decks, or by two to five bulkheads from an engine room, do not require additional noise treatments.

6.6   Auxiliary Equipment

Auxiliary equipment such as pumps, compressors, and purifiers are usually less noisy than main engines. However, an auxiliary machinery room is generally closer to the accommodations than the main machinery space. As such, the potential impact of auxiliary equipment needs to be considered. The largest consideration for other machinery—compressors, pumps, fans, etc.—is whether isolation mounting of the equipment is more effective versus path and receiver treatments for noise abatement. Typically, if only one or two machinery items dominate the received noise in multiple locations, then treatment of the source is most effective. When there are numerous sources and they are well distributed, then receiver treatments are generally the most effective approach. These receiver treatments are generally 'high transmission loss' (HTL) bulkhead or deckhead constructions, increased absorption for the finish surfaces, and floating decks or rooms. To treat the path, damping is the most commonly used method.

Another noise abatement method that is often advantageous is to compactly place or group some equipment on one skid having a common bedplate or skid. This skid can then be isolation mounted, optimising the number and placement of flexible connections.

Pumps are noisy if cavitation occurs, which can happen with an improper pump size or type.

Where possible, rotary (screw or gear) pumps should be selected over piston pumps. Piston pumps usually require vibration isolation mounting and flexible hoses to reduce structure-borne noise.

Path Treatments

7

7.1 Introduction

This chapter discusses treatments that can be used to abate both airborne and structure-borne transmitted noise. The "noise path" is defined as the construction between the source and the receiver space. The receiver space may be a workstation or compartment, where the noise criteria are applicable. It is necessary to distinguish between the structure-borne path and the airborne path.

The structural path starts at the source's foundation and includes all vessel structures up to but excluding the receiver space(s). For high source levels, structure-borne sound can be the dominant noise source and path from machinery spaces to accommodation areas. Acoustic energy is transmitted by elastic waves through structures in many ways, including direct path and multiple flanking paths. The airborne path is usually a significant factor only within the source spaces and in spaces adjacent to the noise source space.

Noise control design options for both paths are discussed below.

7.2 Airborne Noise

The following steps can be taken during the design stage to minimise airborne noise components. One effective approach is to separate machinery spaces from noise-sensitive compartments by applying the following guidance:

- Place compartments that have higher noise criterion between the machinery space and the more noise sensitive compartments. This typically includes storeroom, lockers, change rooms, and in some cases, passageways and toilets/showers,

F. Karkori, *Ship Vibration 3*, Synthesis Lectures on Ocean Systems Engineering,
https://doi.org/10.1007/978-3-031-68078-6_7

- Fan and auxiliary rooms, if located in the superstructure or near noise-sensitive spaces, should be located so as to minimise the common interface area,
- Line shaft, elevators, and stairways should not have direct unimpeded connections between machinery spaces and manned compartments,
- The exhaust casing should be a structure separate from the superstructure containing accommodations, if possible,
- Machinery room intake/exhaust systems should not be located near on-deck stations or other potentially critical areas such as lifeboat stations,
- Fan inlets/outlets, where possible, should face away from noise sensitive spaces such as cabins, hospitals, etc.,
- Accommodations and other noise-sensitive spaces should not be directly over/adjacent to high noise spaces such as the main engine room.
- Use an acoustic enclosure where possible, taking into consideration access needed for maintenance, cooling, and fire control,
- Minimise or avoid openings and pipe/cable penetrations in partitions between source room and receiver room, and
- Weather deck stations may be protected from stack and fan airborne noise with silencers, screens, or by controlling directivity of noise radiation from the pipe stack.

Proper application of these steps will help increase the transmission loss (TL) of the common partition. Several other treatments can also be used to accomplish this increase. The lightest (by mass) treatment is the addition of 50–100 mm (2–4 in.) thick acoustically absorptive materials to cover the structural surface (bulkhead, deck, or deckhead). This material generally consists of fiberglass or mineral wool with a density of 50–80 kg/m^3 (3 to 5 lbs/ft^3). This material can be placed on either the source side or on the receiver room side. The noise reduction from this treatment can be between 3 and 7 dB in the low to mid frequency range and up to 12 dB at higher frequencies.

Alternative noise treatments, such as additional/thicker joiner panels, can achieve approximately the same noise reduction while others, such as the use of a heavier insulation on the order of 160 kg/m^3 (10 lbs/ft^3), can further improve the TL of the bulkhead. Use of high-density mineral wool has also been found to be as effective as the typical fibreglass/limp mass layer/fiberglass treatment and may be easier to install.

7.3   Structure-Borne Noise

The following general design factors can be considered to reduce the influence of structure-borne noise:

- The longer the distance and the greater the number of obstacles between the source and the receiver space, the greater the reduction of structure-borne noise. Obstacles are basically intersections of hull structures, such as connections of decks and bulkheads,
- Stiffer and heavier foundation top plates help reduce vibration amplitude around the foundation, and as a result less energy will spread to remote spaces,
- Structural energy is readily spread through stanchions. If possible, avoid stanchions or do not land stanchions directly on propulsion machinery foundations,
- Hard-connected machinery piping helps facilitate structure-borne noise; this can be avoided by resilient connections of the piping with machinery and structure.

To specifically treat the structure-borne path itself, the application of a damping layer or coating is typically used. Damping treatments have the ability to reduce the vibration level of vibrating plates due to non-recovered loss of energy converted to heat. The effectiveness of damping treatments or coatings depends on the following factors:

- Physical constants of materials such as loss factor of coating, stiffness and weight of the coating, thickness of coatings, coating/substrate thickness ratio, and temperature,
- Location and area of coverage.

Damping treatments may reduce plate vibration levels by approximately 3–10 dB. The lower number (3 dB) references the frequency range of 100–300 Hz, while the higher number (10 dB) references frequency levels of 2,000 Hz and above. However, the actual reduction in the received noise level may be slightly lower. As an added advantage, damping also increases the airborne noise transmission loss, but to a lesser degree than absorptive layers or joiner panels. A 3–5 Db decrease in the airborne transmission may be feasible. Thus, it is important to remember that damping treatments can simultaneously reduce structure-borne and airborne noise components.

Two types of damping coatings are more commonly in use: free surface damping and constrained layer damping (refer to Fig. 7.1, "Unconstrained tile damping applied to hull side").

Typically, the greater the deformation of the damping layer the greater the effectiveness of the damping treatment. A relatively thin metal or composite constrained layer may cause the damping material to work more effectively. A 2–5 dB difference in effectiveness may be found when constrained and unconstrained layer damping is compared for the same type of treatment.

03/23/2005

Fig. 7.1 Unconstrained tile damping applied to hull side

7.4   General Principle for Damping Treatments

The most effective application of a damping treatment is achieved by applying it in a receiver space. In this case, noise will be reduced only in the area of application. Noise reduction per area of coverage is maximised by a local application. If damping tiles are applied on the foundation and around the foundation (tank top, sides) of a critical vibrating source (as discussed above), then structure-borne noise contribution from this source will be reduced throughout the vessel.

However, the overall noise reduction attained will be less than if a local application were applied because of the existence of multiple flanking structure-borne paths.

Careful installation details should be developed for all types of damping treatments, but particularly for damping tiles.

7.5   Machinery Resilient Mounts

The following is design guidance for resilient mounting of machinery:

- Stiffness of the resilient mounts should be much less than the stiffness of the machinery foot and the foundation,
- The resilient mounts should be properly loaded. Do not exceed vendor-recommended limits,
- There should be no rigid structural connection between the mounted machinery and the structure,
- The natural frequency of the mounted machine in all six degrees of freedom should be at least a factor of two below the lowest frequency rate of significant excitation (e.g., rotation rate, blade rate, and firing rate) by the vibration source,
- Resilient mounts should be located over hard spots of the foundation. Hard spots are created by the use of gussets between the top plate and foundation floors,
- All resilient mounts should be captive, either inherently or by the addition of external snubbers,
- Snubbers should be used to limit displacement under impact loads and to prevent metal-to-metal contact
- Painting of the insulation mounts should be prohibited,
- Resilient mounts should be located for easy visual inspection, and
- The rubber mounts are subject to aging; they should be replaced periodically in accordance with manufacturer schedule.

7.6   Flexible Machinery Connections

For resiliently-mounted systems, it is necessary to avoid flanking paths (a solid or hard connection bridging the mounting system). Ducts, piping, cables, shafting, and exhaust systems that are attached to both the structure and to resiliently-mounted machinery are the obvious flanking paths.

To help prevent such connections from becoming structurally flanked, it is best if they contain flexible couplers capable of providing structure-borne noise transmission losses comparable to the losses across the machinery mounts. The following guidance is related to the installation of flexible couplers:

- Hoses on pipe runs should be flexible in at least two different planes. The total configuration should consist of two sections of straight flexible piping (the minimum free length in each leg of any of the configurations should be 18 cm (7 in.) plus 4 times the pipe diameter) or by inline double arch flexible hose (refer to Fig. 7.2, "90° flexible hose connection"),

Fig. 7.2 90° flexible hose
connection

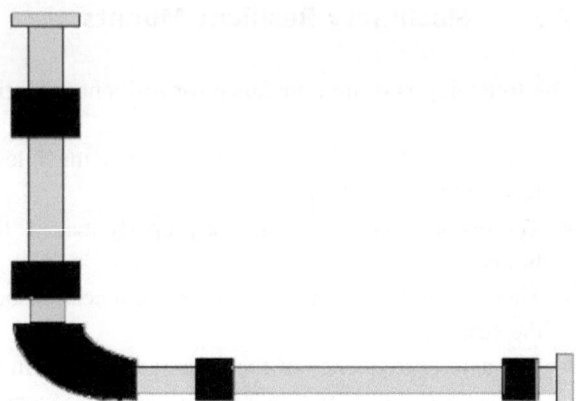

- Rigid fixtures between sections of flexible piping should be supported by resilient hangers,
- A minimum bend radius (not less than that specified by the manufacturer) should be used for cable attachments or a minimum of 36 cm (14 in.) plus 10 times the cable diameter of free cable length with approximately 8 cm (3 in.) of slack should be allowed between the attachment to the equipment and the first cable hanger. Finally, cables and conduits greater than 15 cm (6 in.) in diameter should be supported by resilient pipe hangers at the first two attachment points away from the vibration mounted equipment,
- Resilient pipe hangers should be attached to heavy, rigid parts of the ship structure for the isolation treatment to be most effective,
- Resilient pipe hangers should have the same natural frequency as the mounts supporting the machinery to which the piping is connected, and
- Exhaust systems should be resiliently attached to rigid ship structures. Care should be taken so that the exhaust system does not resonate and that the proper loads are placed on the exhaust mounts, accounting for cold and hot loads (similarly for ducted intakes).

7.7   Flexible Shaft Coupling

The flexible shaft coupling has two purposes: to attenuate vibration from the source into a shaft that is connected to the ship structure by shaft bearings and to compensate for misalignment between shaft's connecting pieces of equipment that are not rigidly attached to the same sub-base or foundation.

The selection of a particular type of coupling is usually determined by the load carried by the shaft, the angular and parallel misalignment that must be anticipated, and the variation in the axial separation between the two coupled shaft ends. Specifying vibration-isolation performance requirements of flexible shaft couplings is generally not possible. However, couplings sufficiently flexible to accommodate the relative movement between shaft ends usually provide adequate isolation to avoid structural flanking of resiliently-mounted vibration sources.

7.8   Flexible HVAC Duct Connections

Resiliently-mounted elements within a duct system should be connected to non-resilient parts of the system by flexible duct connections. Ducts containing fans, whether resiliently-mounted or not, should be flexibly connected to their associated duct work. These flexible connections should, as a minimum, have sufficient flexibility to allow full and free motion of the duct under all ship operating conditions. These connections should be non-metallic where system requirements allow.

Rat guards, if required, should not impede the free motion of the flexible connections. Care should be taken in their design and installation so that there is no metal-to-metal contact during maximum excursion of the equipment under ship slamming conditions.

Where breakout noise (the escape of noise from within the duct into the noise-sensitive compartment) could adversely affect a compartment's airborne noise level, special attention to the design and treatment of the flexible connection is required. Flexible duct connections should not be used to correct for misalignment. Duct work and resiliently-mounted equipment should be aligned within 3 mm per 25 mm of flexible duct length as a minimum.

7.9   Fluid Systems

Fluid systems should be designed with the lowest flow velocities consistent with other system requirements. A maximum velocity of 1.8 m/s (6 ft/sec) should be used as the design parameter for all water systems, with 4 m/s (13 ft/sec) being the maximum flow velocity allowable.

All piping and components should be as clean and smooth internally as possible. All systems should be thoroughly flushed and cleaned to minimise internal roughness. Abrupt changes in flow direction and pressure should be avoided, and all mechanical parts exposed to the flow should be as secure and rigid as possible. Avoid ill-fitting gaskets and other flow discontinuities. The radius of pipe bends should be a minimum of five (5) diameters for low noise generation.

Orifice plates in a piping system should be avoided wherever possible. Single, high pressure drop orifice plates should not be used for system balancing. Use multiple plates in series at mismatched distances. Follow orifice plates with straight pipe for at least seven (7) diameters. An alternative to single-hole orifices is multiple-hole orifice plates.

Globe and angle valves are known to cause noise problems and should only be used where it is impractical to use any other system and/or on systems that do not affect the acoustic environment of the vessel. Throttling valves can cause severe noise problems under conditions where local cavitation can occur. If throttling is necessary in a system, the quietest method is the use of multiple pressure drops.

HVAC Treatments

8

8.1 Introduction

Central ventilation and air-conditioning systems may have noisy elements such as compressors, chillers, and fans. If located in an engine room, the noise from these units is masked by noisier sources. If they are located in a special (fan) room, this room may be treated as a machinery room, where compressors and fans are the sources of airborne and structure-borne noise. Similarly, as previously discussed, machinery noise treatments can be applied to reduce noise in a fan room and in adjacent rooms (resilient mounts, absorptive insulation, damping, etc.).

Central HVAC systems usually have a well-developed duct system with many branches and turns.

Depending on fan noise level and air flow speed, the noise levels at the terminal (louver) may be significant. The application of good 'acoustic' design practices discussed later in this chapter can help control HVAC-induced noise. Generally, it is recommended to use low noise fans and/or low-speed flow rates and to conduct flow and fan noise predictions.

8.2 Design verses "Add-On" Features

Several of these features are useful as an "add-on" or supplementary noise control treatment to an HVAC system that has been acoustically designed, while others are more or less artifacts of the system's requirement to distribute air. The placement of flow divisions, turns, and openings is controlled more by HVAC system requirements than by acoustic requirements. Energy dissipation at duct walls, plena, and silencers can all be designed into the system as "added-on" when noise control becomes necessary.

© The Author(s), under exclusive license to Springer Nature Switzerland AG 2025   57
F. Karkori, *Ship Vibration 3*, Synthesis Lectures on Ocean Systems Engineering,
https://doi.org/10.1007/978-3-031-68078-6_8

8.3   Fans and Individual Fan Coils

Addressing an individual compartment's fan coil units and central air-conditioning systems requires a different approach to noise control than an entire HVAC system. As with pumps, a fan needs to operate near peak efficiency in order to reach its lowest noise level. Fans operating outside their design curve will have higher noise levels than fans operating at peak efficiency.

All ventilation fans should be resiliently mounted and all connections to the fan, including ducting and electrical cables, should be flexible and at least as effective as the vibration isolators. If it is possible, fans should not be attached to bulkheads, decks, or overheads that form the boundaries of noise-sensitive spaces. Fan foundations should not be cantilevered from a bulkhead.

As a rule, the airborne noise level from a fan coil, listed in the acoustic data for the unit, should be 5 dB lower than the compartment's criteria. This acoustic data is typically provided by the manufacturer.

8.4   Duct Construction

Airborne noise inside the duct that is transmitted through the duct wall (breakout noise) into a noise-sensitive space can be as important as the noise transmitted through the diffuser or terminal into the space. High noise levels at low frequencies that exist in most HVAC systems are usually difficult to prevent from transmitting through duct walls.

Generally, round or oval ducts provide higher losses for noise transmitted through the duct walls than rectangular or square ducts. This is particularly true at the low to mid frequencies. In addition, circular ducts attenuate noise moving down the duct better than rectangular ducts.

Finally, low frequency rumble caused by turbulence is less likely to occur in circular ducts than in square ducts because circular ducts are usually more rigid. For these reasons, round or oval ducts should be used wherever possible in HVAC systems for ducts passing through noise-sensitive space.

The fan discharge should be straight for a distance of at least three major duct widths to avoid turbulence. Noise will significantly increase due to turbulence from elbows too close to the discharge (or inlet).

Table 8.1 Length of duct proceeding fitting

Turn size(degrees)	Diameters of straight duct
30	2
45	3
60	4
90	5

Fan inlet air flow should be straight for at least three duct diameters. Turns closer than three diameters cause air flow turbulence which increases noise. Five to ten diameters of straight duct is required for turbulence to die out and flow to equalize. If the air flow does not become smooth before the next fitting or terminal device, the flow noise generated in the next element will increase.

Square turns, mitred elbows and zero radius elbows should generally be avoided. Transforming sections should be symmetrical. Avoid using abrupt changes in cross-section or duct turns. Noise increases when expansion in a transition section exceeds a 7° angle from the straight-ahead.

The use of vane turns is advised when the turn is within five duct diameters of a fan. If the bend is more than five diameters from the fan, regular elbows with ample radii should be used as they cause less pressure drop in the flow.

The area of the duct that engages a fan should be greater than or equal to the active area of the fan.

Also, fan intakes should be kept symmetrical with respect to the fan.

The inside of the ducts should be as smooth as possible to avoid turbulent flow. Seams should be faired, and protruding objects avoided. Improperly fitted gaskets and flex couplings should be avoided as well. Leading edges of dampers, splitters, and deflectors should be rounded or folded back.

Use the following straight length of duct preceding the fitting, when using vaned or un-vaned elbows (refer to Table 8.1, "Length of duct proceeding fitting").

To eliminate sources of noise, adjustable splitters should be avoided. Orifices are preferred for balancing of systems. Splitters, if used, should be designed to be permanently and rigidly fixed in position after the systems are balanced.

Rat guards on flexible couplings should not bridge the flexible coupling. The guard should be rigidly attached to only one end of the ductwork.

8.5   Treatment of Ducts

The attenuation of duct-borne sound transmission from the major source (e.g., fans, airflow, etc.) to the receiver space air diffuser or terminal should be addressed and (louver) is affected by:

- Acoustic energy dissipation at duct wall surfaces,
- Branches where the flow divides,
- Turns where the flow changes direction by greater than 30°,
- Plena,
- Duct silencers,
- Duct openings.

The attenuation of noise propagating through straight duct runs is a function of the length of the run and the amount of absorption along the wall surfaces. A 3 m (10 ft) length of 25 cm (10 in.) circular duct with 25 mm (1 in.) of internal lining may provide an attenuation of 14 dB in the mid to high frequencies. An unlined duct of the same dimensions may not provide any measurable attenuation. As with most other absorption treatments, lined ducts may not be effective below 250 Hz. The average attenuation, in decibels per meter, for ducts with 25 mm (1 in.) internal acoustic lining is around 3 dB/m (1 dB/ft) in the mid frequency range.

The airborne transmission path in straight duct runs is often flanked by structure-borne paths in duct walls, since the airborne noise in the duct excites vibrations of the duct walls. The noise is transmitted along the duct wall and radiated downstream as airborne noise in the duct. Unless the structure-borne path is broken with flexible duct couplings, the maximum attenuation of in-duct airborne noise that can be expected is generally about 25 dB for the mid frequencies. Therefore, without flexible duct couplings, the length of lined duct should not exceed 6 m (20 ft). 9 m (30 ft) of lined duct would generally not be more effective than 6 m (20 ft) unless structure-borne flanking in the duct wall is treated. In general, most HVAC systems should have at least 3 m (10 ft) of internally-lined duct upstream and downstream of all fans to reduce fan in-duct noise transmission. Most ventilation systems with flows greater than 5 m3/s (10,000 cfm) should have internally-lined ducts with a thickness of at least 50 mm (2 in.) for a distance of approximately 4.5 m (15 ft) to reduce fan airborne noise transmission in ducts.

If additional attenuation in the duct is necessary, duct turns should be lined for a distance of at least three duct diameters before and after the turn. This can also help reduce cross-talk between rooms served by the same duct system, which can be further avoided by providing offsets. With these measures in place, there may be an additional 3 to 6 dB of attenuation at high frequencies, depending on the size of the duct. Typically, the larger the diameter of the duct, the more effective the lined turn is.

8.6 HVAC Silencers and Plena

When the pressure drop can be tolerated and space permits, HVAC silencers can be used to provide significant noise attenuation for systems having noise excesses. Another noise control treatment of in-duct airborne noise transmissions is to line plena with acoustic absorptive material.

Lining the plena creates a "silencer" in the system. The advantage of a lined plenum is that the attenuation characteristics are high at low to mid frequencies, where lined duct attenuation is low at these frequencies. Careful design of the plena and the placement of baffles, if necessary, can lead to the removal of discrete or narrowband tones caused by many fans. The plena chamber(s) can be designed to have a damped acoustic resonance at the offending tone of the fan. The attenuation offered by a lined plenum chamber can be as high as 30 dB, and the attenuation usually has a strong-frequency dependence.

A three-pass plenum (two baffles in the flow path) can provide up to 50 dB of attenuation. The chief disadvantage of a lined plenum is its high-pressure loss. If plena exist in a system already, lining the chamber walls can provide significant noise attenuation (Beranek, 1971).

8.7 Lagging Treatments

Lagging treatments can be used to reduce the noise radiated from the surface of vibrating duct systems. Most lagging treatments consist of a layer of fibreglass covered by a layer of impervious material such as a sheet of aluminium, or leaded vinyl. The effectiveness of such a cladding treatment is small at low frequency. However, it can reach 20 dB at high frequencies.

8.8 Flow Rate

Air flow velocities should be kept as low as possible. For design guidance, the target flow rates should be no higher than 5 m/s (1,000 ft/min.) for noise-sensitive compartments, and never exceed 12.7 m/s (2,500 ft/min) for machinery ventilation systems. Higher velocities require extensive treatment, particularly near each diffuser terminal. Table 8.2, "Maximum airflow for various ratings" provides recommended maximum airflow velocity for various installations.

Table 8.2 Maximum airflow for various noise ratings

Duct type	Compartment noise limit dB(a)	Maximum airflow in duct fpm
Square turn	75	2400
	65	1600
	55	900
Radius square turn w/short vanes	75	2900
	65	2000
	55	1000
Square turn, long vanes	75	3200
	65	2500
	55	1200
Radius turn w/long vanes	75	3300
	65	2700
	55	1200

9.1   Introduction

In general, noise control in a receiver space may be needed if:

- Noise controls applied to the source and noise path are not sufficient,
- The source/path treatments are too expansive to reach the acoustic goal, or
- Given the distribution of sources, it is more effective to implement receiver treatments.

Once it has been determined that receiver treatments are required, two kinds of receiver spaces should be distinguished: adjacent to the source room and remote from the source room. These two groups differ with contribution of airborne and structure-borne noise. An adjacent room has an airborne noise component that is usually noise transmitted through a partition between the source and receiver rooms, while the noise levels in the remote rooms are purely structure-borne.

A further breakdown within the room itself is also necessary, making sure to distinguish between the vibration reduction of the room's boundary and reduction of the ability of the surfaces to radiate noise.

9.2   Damping Treatments

Damping treatments discussed in Chap. 8 used to reduce the vibration levels of the compartment's boundaries, along with acoustically absorptive material in the overhead (to be absorptive, the pan supporting any insulation needs to be perforated in order for the sound to be 'absorbed' by the insulation) are primary means to reduce noise levels in all noise-sensitive compartments. This works by increasing the room's ability to absorb

F. Karkori, *Ship Vibration 3*, Synthesis Lectures on Ocean Systems Engineering,
https://doi.org/10.1007/978-3-031-68078-6_9

airborne noise energy and so reducing the levels of unwanted noise. To achieve the best results, the acoustically absorptive material should be placed on the internal 'finished' surface(s) of the compartment. Note that the insulation is not effective for absorption if it is placed behind a solid sheathing or joiner panel but will increase airborne transmission loss if placed in this void. The expected noise reduction associated with increasing the absorption in a compartment may be on the order of 2–4 dB (increasing the absorption in a machinery compartment will not affect the noise in the vicinity of a machinery item since the absorption only impacts the 'reverberant noise').

9.3   Other Receiver Treatments

Another one of the more effective treatments used to reduce the noise in the receiver space is a floating floor (refer to Fig. 9.1, "Floating floor treatment"). With this treatment, the finish floor (sole) is resiliently mounted on the structural deck. The resilient element may be a uniformly distributed layer, such as a mineral wool layer.

Yet another option is the use of individual resilient mounts placed between the sole and the structural deck. To be fully effective, it is recommended that joiner panels be installed on top of the floating floor without using a hard connection between the joiner panel and hull structure. The sole may be covered with a damping coating in order to increase the overall effectiveness. Noise reduction on the order of 10–15 dB may be expected from a standard floating floor.

Fig. 9.1 Floating floor treatment

Propulsion Treatments

10

10.1 Introduction

The most effective way to minimise propulsor noise is through careful design. The designer's goal should be to increase the cavitation inception speed (rotation rate at which cavitation begins). The following measures can help in increasing the cavitation inception speed:

- Uniform flow in the propulsor disk,
- Skewed blades of the propulsor,
- Increasing propeller diameter thereby reducing the needed rpm,
- Increasing number of blades.

A rule-of-thumb for tip-hull and leading edge-strut/rudder clearance is nominally 20% of the propeller diameter. All other things being equal, the hull vibration response will decrease with increasing clearances.

Vibration of hull structure around propulsors may be reduced by using a damping coating or by increasing hull plate mass and stiffness.

10.2 Piping System Treatments

As with the HVAC system, piping systems produce noise at the pump, which should be treated like a machinery item, as discussed in this chapter. Additionally, consideration should be given to the piping system itself. These systems can be designed to be quieter by reducing the flow speed and avoiding sharp bends.

© The Author(s), under exclusive license to Springer Nature Switzerland AG 2025
F. Karkori, *Ship Vibration 3*, Synthesis Lectures on Ocean Systems Engineering,
https://doi.org/10.1007/978-3-031-68078-6_10

Fig. 10.1 Typical resilient
pipe support

When treatments are needed, there are limited choices. Various hydraulic silencers exist which can reduce the pump pressure pulsations getting into the fluid and hence the pipe wall. These can either be tuned or broadband silencers. A flexible hose can attenuate both the fluid-borne and structure-borne noise from the pump entering the piping system. Finally, for piping systems with high vibrations, resilient attachment points can be used, refer to Fig. 10.1, "Typical resilient pipe support".

10.3   Treatment Summary

Table 10.1, "List of treatments and effectiveness" summarises various treatments, their effectiveness against airborne (AB), structure-borne (PSB) and secondary structure-borne (SSB) noise paths. A cost estimate is also provided.

Active noise controls are not typically applicable to control shipboard noise except for use in HVAC ducting and exhausts.

10.4   Fire Safety of Treatments

Materials used for noise control treatments are to comply with applicable fire safety requirements. Where structural fire protection is required, an efficient design may use materials having required fire safety approvals and acoustic treatment characteristics.

In locations with thermal or acoustic treatment requirements, the applied material is usually considered as comfort insulation. Comfort insulation may be required to obtain approvals related to flame-spread and or smoke toxicity depending on the application. Requirements will be subject to Flag State regulation or Safety of Life at Sea (SOLAS) requirements. Fire safety requirements may require that only non-combustible materials are used.

Table 10.1 List of treatments and effectiveness

Noise control treatment	Description	AB	PSB	SSB	Cost estimate
Low noise equipment	Selection of equipment that by its design or quality are lower noise and/or vibration	5–10	5–10	5–10	Est. 1 × to 5 × of cost of standard equipment
Machinery vibration isolation (single stage)	Reduction of vibration by mechanically isolating machinery from supporting structure	0	10–20	0	$20-$1,000 per vibration mount
Machinery vibration isolation (double stage)	Use of two layers of vibration isolation mounts under machinery with seismic based between the machinery and the ship's foundation	0	20–40	0	Single stage costs plus seismic base ($10-$20 k)
Damping tiles	Reduces vibration energy in structures	3	5–10	5–10	Material: $5-$15 per square foot
	Used on stiffened plating near machinery sources, plating adjacent to water, and locations in-between				Additional costs apply
Spray-on damping	Reduces vibration energy in structures	2	3–8	3–8	$15-$30 per square foot
					Additional costs apply
	Used on stiffened plating near machinery sources, plating adjacent to water, and locations in-between				
Ballast-Crete	Pumpable material used as ballast but can also be used as damping in voids and tanks	0	5+	5+	N/A
Air masking systems	Air bubble curtain is used to shield vessel hull from rest of ocean	>10	>10	>10	Est. $20,000 to $50,000
Decoupling materials	A decoupling material is applied to the exterior (wet side) plating in order to reduce the radiation efficiency of the structure	>10	>10	>10	Development R&D

It is recommended that the acoustic treatment plans be completely cross-referenced with the structural fire protection plans. Acoustic treatments details should include call-outs of the fire rating approvals of the specified materials and note the required ratings according to the structural fire plan.

Damping treatments often include visco-elastic materials. Many of the currently available visco-elastic damping treatments are not approved as non-combustible materials and are not allowed where only non-combustible constructions are required. Some damping treatments have approvals permitted as low flame spread surface coverings. The regulations or approval certificates may limit the permissible thickness of treatments approved as surface coatings. In some applications, the approvals prohibit the materials to be covered. In this case, the treatment is approved as an exposed facing.

Several useful damping treatments are approved as deck covering systems. Application of these treatments are limited to deck covering, unless specifically allowed for application to bulkheads and other non-deck covering surfaces. The fire safety characteristics of marine acoustic treatments are a very important aspect of the design. Some commercial literature for acoustic treatments and damping describes the materials as "approved".

Not every "approval" certificate is appropriate for all applications. The buyer and designer are urged to use caution and to double check that the specified material is truly acceptable with regard to the fire safety requirements of the project. Safety of life at sea may depend on the fire safety of the acoustic materials.

Implementation 11

11.1 General

This brief chapter addresses the implementation of the noise control efforts on board the vessel. Once the treatments have been selected and detailed, it is just as important to follow up with good quality assurance programme compliance testing. The publication, *Applying Physical Ergonomics to Modern Ship Design* (Olsen, 2024) provides detailed requirements on the quality assurance procedures and testing procedures to be submitted to Class for consideration.

11.2 Quality Assurance

It is absolutely necessary to implement a good quality assurance programme, including a thorough review of drawings and the actual implementation of the treatments onboard the vessel. These steps provide insurance that the time, effort, and cost put into the design effort pays off.

11.2.1 Drawing Review

Drawing reviews are an integral part of the noise control effort, providing needed support and guidance. As the noise control treatment designs evolve, drawings and other design information needs to be reviewed by one familiar with the design of treatments to verify that the treatment design adequately reflects acoustic considerations. Results of

F. Karkori, *Ship Vibration 3*, Synthesis Lectures on Ocean Systems Engineering,
https://doi.org/10.1007/978-3-031-68078-6_11

these reviews need to be documented and appropriate review remarks made on noise critical drawings. Drawings need to be revised to incorporate any suggested changes to the treatment details.

11.2.2 Construction Inspection

The construction of noise control treatment installations should be thoroughly inspected. This includes the inspection of noise-critical machinery and treatments so that the control treatment installations are not compromised. The inspections should primarily address the fabrication and installation of acoustic insulation materials, damping, high transmission loss materials, double decks/bulkheads, enclosures, isolation-mounted equipment, silencers, and other critical treatments. The performance of noise control treatments is ultimately dependent on the quality of the implementation. Seemingly trivial deviations from the detail design or inadvertent errors due to unfamiliarity with noise control treatment materials and constructions may compromise acoustic performance.

Comments and action items should document and identify constructions and installations that are judged to adversely impact noise performance.

11.2.3 Trials

Example acoustic test procedures can be found in *Applying Physical Ergonomics to Modern Ship Design* (Olsen, 2024). This includes guidance on test plans, personnel, conditions, data acquisition and instrumentation, data analysis, test schedule, requirements, and reports.

11.2.4 Material Selection

Materials selected for installation need to meet regulatory requirements. The most critical of these requirements pertain to fire, smoke, and toxicity. Some materials like damping tiles, for instance, can be used in machinery spaces but not in accommodations. However, the latest spray-on damping materials are approved for use in both machinery spaces and accommodations.

It is recommended that material vendors with direct experience in the marine industry be approached before considering vendors with only 'industrial' experience. The usual 'installation' standards used by the shipyard and their prior experience in noise control also bears consideration in selecting noise abatement approaches and materials.

11.3   Summary

Noise objectives can be met using the processes discussed throughout this book. The first step is developing the appropriate noise control plan. Next, design for low noise levels should be used to minimise the need for add-on treatments. A detailed noise analysis will reveal the critical sources and acoustic transmission paths. With this detailed understanding of the potential problem areas, the selection of appropriate and optimal treatments can be conducted with confidence. These treatments need to be incorporated into the detailed design and installed in the vessel. At this point, it is critical that the treatments be inspected to make sure they are installed properly without any shorts or other elements that would compromise their performance.

Noise analysis during the design stage, engineering of noise control, and participation in construction control and noise measurements should be performed by companies or personnel with a noise control background and experience.

Correction to: Ship Vibration 3

Correction to:
F. Karkori, *Ship Vibration 3*, Synthesis Lectures on Ocean Systems Engineering,
https://doi.org/10.1007/978-3-031-68078-6

This book contains overlap in text with the previously published content [1] that was inadvertently omitted. The authors failed to attribute the reference [1]. The authors have now obtained permission to re-use this content from the American Bureau of Shipping.

Where [1] is: American Bureau of Shipping (2024), Rules and Guides https://ww2.eagle.org/en/rules-and-resources/rules-and-guides.html

The updated version of this book can be found at
https://doi.org/10.1007/978-3-031-68078-6

Glossary

Acceleration A vector that specifies the time rate of change of velocity (units of m/ s2). The acceleration of the vibratory motion of a structure can be specified in terms of the peak, average, or root-mean-square (rms) magnitude of the acceleration in a given direction. In this document, the acceleration levels are given in terms of the rms acceleration amplitude.

Acceleration level LA, in dB, of a vibrating body is 20 times the logarithm to the base 10 of the ratio of the acceleration to a reference acceleration (standard reference acceleration level is 10-3 cm/s2).

$$LA = 20\log(a/aref) = 10\log(a/aref)2.$$

Acoustic For the purposes of this document, "acoustic" refers to both noise and vibration related phenomena.

Acoustic absorption The change of sound energy into some other form, usually heat, in passing through a medium or on striking and being reflected from a surface.

Acoustic power See "sound power" or "sound power level".

Airborne sound (or noise) Sound or noise that is transmitted through air or by means of paths in air. Since air cannot support shear, only pressure waves with longitudinal displacement can be transmitted through air.

Amplitude The value of vibratory response due to a steady state periodic process. Amplitude may be measured as displacement, velocity or acceleration and may be presented as a peak level or a root mean square (rms) level.

A-weighted sound pressure level The magnitude of a sound, expressed in decibels (i.e., 20 micropascals); the various frequency components are adjusted according to the A-weighted values given in IEC 61672.1 (2004) in order to account for the frequency response characteristics of the human ear. The symbol is LA, and the unit is dB(A). The measurement LAeq is an equivalent continuous A-weighted sound pressure level, measured over a period of time.

F. Karkori, *Ship Vibration 3*, Synthesis Lectures on Ocean Systems Engineering,
https://doi.org/10.1007/978-3-031-68078-6

Bending mode Mode of vibration in which cross-sections of a beam, shaft, or structure undergoes translation and rotation. Type of translational mode usually found in slender structures with evenly distributed mass and stiffness.

Calibration checks Field calibration of a measuring instrument conducted before and after a field test, using a reference calibrated signal or through zero calibration.

Comfort The ability of the crew to use a space for its intended purpose with minimal interference or annoyance from noise.

Crest factor The ratio of the peak value to the root-mean-square (rms) value of the acceleration after it has been frequency weighted by the appropriate frequency weighting network.

Crest Factor = weighted peak acceleration weighted rms acceleration

Damping The dissipation of energy with time or distance. In this document, damping generally refers to dissipation of vibrating energy in structures.

Decibel (dB) A dimensionless unit of measure of the ratio of two quantities, P1 and P2, each of which is equal to or proportional to power. Ten times the logarithm to the base 10 of the ratio, P1/P2, has the dimensions of dB.

Direct sound field A sound field in which energy is flowing outward from the source without interference from surrounding surfaces. The sound field very close to a source, even in a reverberant room, is a direct field. Sound fields outdoors are direct fields at all distances from the source and are referred to as "free sound fields" or "free fields".

Duration Is represented by the length of exposure to sound.

Dynamic positioning A system to automatically maintain a workboat's position and heading by controlling propellers and/or thrusters. Dynamic positioning can maintain a position to a fixed point over the bottom, or in relation to a moving object (such as another vessel). It can also be used to position the vessel at a favorable angle towards wind, waves, and current.

Equivalent continuous a-weighted sound pressure level The A-weighted sound pressure level of a notional steady sound, over a certain time interval, which would have the same acoustic energy as the variable-loudness real sound under consideration, over that same time interval. The symbol is LAeq; the unit is dB(A).

Excitation A time-dependent stimulus (force or displacement) that produces vibration. Excitation may be transient, random, and periodic. A steady-state periodic excitation, like that produced by propellers or propulsion engine, is of interest in these Guidance Notes.

Exciting frequency Is the number of cycles of the excitation completed during a given time unit. It may be shaft rate (number of a shaft revolution per second), or blade rate (shaft rate multiplied by number of propeller blades). Under steady state condition, the frequency of vibration is always equal to the exciting frequency. An exciting force at any frequency potentially can excite all modes of the elastic body.

However, amplitude of vibration depends on how close the excitation frequency is to any of the natural frequencies. The modes where the natural frequency is close to the exciting frequency will be most active and dominate in vibration response.

Flanking path A circuitous path by which sound is transmitted around the main transmission path (e.g., paths by which sound is transmitted from machine to foundation bypassing transmission through resilient mounts or gaps around the perimeter of the panel or cracks and openings in the panel through which sound bypasses transmission through the panel).

Free sound field (free field) A sound field in a homogeneous, isotropic medium free from boundaries. The sound pressure level decreases by 6 dB per doubling of distance from the source.

Frequency The number of complete cycles of a periodic process occurring per unit time. Frequency is expressed in Hertz (Hz) which corresponds to the number of cycles observed-per-second.

Frequency band An interval of the frequency spectrum defined between upper and lower "cut-off" frequencies. The band may be described in terms of these two frequencies, by the width of the band, and by the geometric mean frequency of the upper and lower cut-off frequencies (e.g., "an octave band centered at 500 Hz").

Frequency weighting A transfer function used to modify a signal according to a required dependence on vibration frequency.

• In human response to vibration, various frequency weightings have been defined in order to reflect known or hypothesized relationships between vibration frequency and human response.

• The frequency weighting used to evaluate whole-body vibration in these Guidance Notes is Wm (whole-body) for all three axes (x, y, and z), in accordance with ISO 6954.

Hard mounting Rigid attachment of a machine to its sub-base or foundation.

Harmonic (1) An integral multiple of a given frequency; a sinusoidal component of a periodic wave. (2) A signal having a frequency that is a harmonic (sense 1) of the fundamental frequency.

Inception speed Vessel speed at which a propeller starts to cavitate.

Intensity of noise Is represented by the loudness of the sound.

Isolation The coupling of a vibrating structure (e.g., machine) to another structure (e.g., foundation or hull) by means of resilient or compliant supports that prevent the transmission of the vibration from the vibrating structure to the coupled structure.

Level In acoustics, the level of a quantity is the logarithm of the ratio of that quantity to a reference quantity of the same kind. The base of the logarithm, the reference quantity, and the kind of level needs to be specified. Examples are sound power levels, sound pressure levels, and acceleration levels. Levels are always expressed in decibels (dB).

Manned space Any space where a seafarer may be present for twenty (20) minutes or longer at one time during normal, routine daily activities. Such spaces would include working or living spaces.

Mode and mode shape A predetermined distribution of vibration amplitude that occurs when an elastic body vibrates freely at any of its natural frequencies. This distribution is called a 'mode'. These modes are enumerated, with the lowest mode number corresponding to the lowest natural frequency. Each mode has a specific shape of amplitude distribution with nodes (null amplitude points) and maximums of vibration.

Multi-axis acceleration value The Multi-Axis Acceleration Value is calculated from the root-sums-of-squares of the weighted rms acceleration values in each axis (axw, ayw and azw) at the measurement point using the following expression:

aw = axw

2 + ayw

2 + azw

2

where axw, ayw and azw are the weighted rms acceleration values measured in the x-, y- and z-axes, respectively.

Multi-axis vibration Mechanical vibration or shock acting in more than one (1) direction simultaneously.

Natural frequency Is the frequency at which structures vibrate freely after a single impulse or impact. A mass-spring system has only one natural frequency. An elastic body like a ship hull (or parts of the hull) has a multitude of natural frequencies. Usually, only the lower natural frequencies are of practical significance.

Noise Any undesired sound or any erratic, intermittent, or statistically random oscillation.

Noise level Noise levels in this document refer to the sound pressure levels that describe a particular noise environment.

Noise reduction The difference (in decibels) between the sound pressure levels on different sides of a boundary construction is defined as the noise reduction (NR). Unlike transmission loss, noise reduction is dependent not only on the intervening structure but also the source and receiver room acoustic properties. Noise reduction can also refer to the number of dB by which sound pressure levels are reduced in a single space or at a fixed location due to introduction of noise control treatments.

Octave The frequency range bounded by upper and lower frequency limits fu and fl, where fu = 2 fl. Octave bands are usually specified by their geometric mean frequency, called the band center frequencies. The standard octave bands covering the audible range are designated by the following center frequencies: 31.5, 63, 125, 250, 500, 1000, 2000, 4000, 8000, and 16,000 Hz. The corresponding upper and lower frequencies are 22/45, 45/89, 89/177, 177/354, 354/708, 708/1416, 1416/2832, 2832/5664, 5664/11,328, 11,328/22,656.

Overall root mean square acceleration Is the root from the arithmetic mean of the squares of the vibration spectrum components.

Peak value The largest deviation of a signal from the arithmetic mean of that signal. The positive peak value is the maximum positive deviation; the negative peak value is the maximum negative deviation.

Period Is the minimum time between consequent repeating amplitudes.

Pure tone Sound wave whose sound pressure at a point varies sinusoidally with time and is characterized by its singleness of pitch. The time interval for one complete cycle of the sinusoidal pressure variation is known as the period.

Radiation Efficiency The radiation efficiency of a vibrating surface is proportional to the acoustic power radiated per unit surface area per unit of mean-square velocity of vibration averaged over the radiating surface. It is the measure of the efficiency with which a given surface converts vibratory energy to acoustic energy.

Reference calibration Calibration of measuring instrument conducted by an accredited Testing and Calibration Laboratory with traceability to a signal or through zero calibration.

Resonance (1) The phenomenon of amplification of a free wave or oscillation of a system by a forced wave or oscillation of exactly equal period. The forced wave may arise from an impressed force upon the system or from a boundary condition. The growth of the resonant amplitude is characteristically linear in time. (2) Of a system in forced oscillation, the condition that exists when any change, however small, in the frequency of excitation causes a decrease in the response of the system.

Reverberant field That part of the radiated sound field where the sound waves reflected from the boundaries of the enclosure are superimposed upon the incident field. The reverberant field may be called a diffuse field if a great many reflected wavetrains cross from all possible directions and the sound-energy density is very nearly uniform throughout the field.

Reverberation room A room in which all boundaries are hard and the sound field is reverberant over nearly the entire room volume with the exception of a small region around the source.

Soft surface A surface that is a good absorber of sound.

Sound A disturbance (i.e., oscillation in pressure) that propagates through an elastic material (air, water) at a speed characteristic of that medium. The physical quantities of most interest are the sound pressure and sound power.

Sound energy The energy that sound waves contribute to a particular medium.

Sound field A region containing sound waves. Sound fields are typically described as reverberant or direct.

Sound pressure The total instantaneous pressure at a point in the presence of a sound wave minus the static pressure at that point.

Sound pressure level SPL, in dB, is 20 times the logarithm to the base 10 of the ratio of the pressure of this sound to a reference pressure. The sound pressure, p, is the root-mean-square value of the instantaneous sound pressure over a time interval at the point under consideration.

Source level In this document, either the sound power level of an airborne noise source or the acceleration level of a structure-borne noise source.

Spectrum (1) The spectrum of a time-varying phenomenon is the description of its resolution into components, each of different frequency and (usually) different amplitude and phase. (2) Also used to signify a continuous range of components, usually wide in extent, within which waves have some specified common characteristic (e.g., "audio-frequency spectrum").

Stationkeeping Maintaining a seagoing vessel in a position relative to other vessels or a fixed point.

Structure-borne sound (or noise) Sound or noise that is transmitted through structures that are capable of supporting shear. Thus, structure-borne sound may be in the form of longitudinal or flexural waves.

Thruster conditions Those conditions when thrusters are used to maintain the vessel's position and heading during stationkeeping.

Transit conditions Those conditions where the vessel is transitioning (moving) from one location to another.

Transmission loss The sound transmission loss (TL) of a partition, in dB, is the difference between incident sound intensity level and transmitted sound intensity level. It can be expressed as 10 times the logarithm to the base 10 of the reciprocal of the transmission coefficient, τ.

Turn A change of direction in a passageway or duct where the angle of direction change is at least 30 degrees.

Vibration The variation with time of the magnitude of a quantity which is descriptive of the motion or position of a mechanical system, when the magnitude is alternately greater and smaller than some average value. Also referred to as structure-borne noise.

Vibration level Measure of the amplitude of vibration.

Vibration spectrum Is usually the acceleration squared as a function of frequency.

Wave A disturbance which is propagated in a medium in such a manner that at any point in the medium the quantity serving as a measure of the disturbance, such as pressure or particle displacement, is a function of the time, while at any instant in time the displacement at a point is a function of the position of the point. (Excerpt from American Standard Acoustic Terminology).

Weighted root-mean-square acceleration value (aw) The weighted root-mean-square (rms) acceleration, aw, in meters-per-second squared, is defined by the expression:

where aw(t) is the weighted acceleration as a function of time in meters-per-second squared (m/s2) and T is the duration of the measurement in seconds.

Weighted spectrum Is the spectrum corrected by some filtering values, depending on frequency; this correction accounts for the sensitivity of a human body to vibration in different axes.

Whole-body vibration Mechanical vibration (or shock) transmitted to the human body as a whole. Wholebody vibration is often due to the vibration of a surface supporting the body.